GEOTECHNICAL SPECIAL PUBLICATION NO. 74

Guidelines of Engineering Practice For Braced and Tied-Back Excavations

By the Committee on Earth Retaining Structures
of The Geo-Institute
of The American Society of Civil Engineers

W0082523

GEO INSTITUTE

Published by

ASCE American Society
of Civil Engineers

1801 Alexander Bell Drive
Reston, VA 20191-4400

Abstract:
When an excavation is made in soil or rock, materials adjacent to the cut face lose their natural lateral support. It then becomes necessary to support the excavation by using braces and tiebacks. These will provide the lateral restraint needed for safety. These guidelines have been prepared to aid civil engineers in making the essential decisions regarding construction safety and excavation activities. They are also intended to provide information and conceptual orientation for owners, funding agencies, contractors, and engineering and architectural staff.

Library of Congress Cataloging-in-Publication Data

American Society of Civil Engineers. Committee on Earth Retaining Structures.
Guidelines of engineering practice for braced and tied-back excavations / by the Committee on Earth Retaining Structures of the Geo-Institute of the American Society of Civil Engineers.
p.   cm.
Includes bibliographical references and index.
ISBN 0-7844-0293-0
1. Excavation. 2. Retaining walls. 3. Shoring and underpinning. I. Title.
TA730.A44 1997   97-36244
624.1'52--dc21   CIP

The material presented in this publication has been prepared in accordance with generally recognized engineering principles and practices, and is for general information only. This information should not be used without first securing competent advice with respect to its suitability for any general or specific application.

The contents of this publication are not intended to be and should not be construed to be a standard of the American Society of Civil Engineers (ASCE) and are not intended for use as a reference in purchase specifications, contracts, regulations, statutes, or any other legal document.

No reference made in this publication to any specific method, product, process or service constitutes or implies an endorsement, recommendation, or warranty thereof by ASCE.

ASCE makes no representation or warranty of any kind, whether express or implied, concerning the accuracy, completeness, suitability, or utility of any information, apparatus, product, or process discussed in this publication, and assumes no liability therefore.

Anyone utilizing this information assumes all liability arising from such use, including but not limited to infringement of any patent or patents.

GEOTECHNICAL SPECIAL PUBLICATIONS

CONTENTS

ix

PREFACE

Work on these guidelines was initiated at the suggestion of the late George W. Sowers, then chairperson of the Earth Retaining Structures Committee of the Geotechnical Engineering Division, at the ASCE *1970 Specialty Conference on Lateral Stresses in the Ground and Design of Earth Retaining Structures.* The work was continued by successive committee chairpersons including: James P. Gould, Harry W. Schnabel, Thomas D. O'Rourke, Donald R. McMahon, Lawrence A. Hansen, and Richard J. Finno. It was originally intended to take the form of an ASCE manual, but the Committee found that consensus could not be reached among members representing different industry interests on many fundamental issues. Therefore, it was finally agreed that guidelines could best express the difference in viewpoints while allowing the user to choose an appropriate approach for a project without the constraints of a manual or a standard. These guidelines are intended for the use of the professional civil engineer involved in the work of a braced or tied-back excavation. This document is not a standard.

Many members of the Earth Retaining Structures Committee contributed comments and concepts to the making of the Guidelines. Those who were particularly helpful in providing text sections and in reviewing text sections include the following members: H. A. Aldrich, R. A. Alperstein, W. J. Armento, R. A. Bell, G. W. Clough, R. D. Darragh, J. P. Gould, L. A. Hansen, A. J. Nicholson, T. D. O'Rourke, H. W. Schnabel, W. L. Schroder, R. vanLeuwen, and D. Weatherby. Text preparation, figure preparation, and final editing of the report were performed by Mary J. S. Roth and Thomas F. Boni.

CHAPTER 1: INTRODUCTION

1.1 Function of a Braced or Tied-Back Excavation and Role of These Guidelines

When an excavation is made in soil or rock the materials adjacent to the cut face lose their natural lateral support. The deeper the excavation and steeper the cut, the greater is the stress unbalance caused by removal of lateral support. If the resulting shear stress exceeds the material's strength the mass can fail by sliding. Facilities or structures outside the excavation can be damaged, work within the excavation disrupted and personnel endangered. When it is impractical to provide slopes flat enough for safety, it becomes necessary to support the excavation. Braces and tiebacks replace the lateral restraint that was provided by material which was removed.

Excavation work, especially in urban areas, involves decisions about complex issues, such as construction methods, support loads, base stability, ground deformation, groundwater control, and influence on adjacent property. These guidelines have been prepared to aid the civil engineer in making such decisions. They also are intended to provide information and conceptual orientation for owners, funding agencies, contractors, and engineering and architectural staff.

The guidelines deal exclusively with braced and tied-back excavations utilized in construction of buildings or other civil engineering structures. Not included are the special cases of temporary support of tunnels or underground chambers, mines, cofferdams placed in open water, excavation schemes for quarries, permanent sloped open cuts such as for highways or site grading and problems of rock slope stability.

The guidelines do not cover trench support placed according to code requirements or empirical rules without engineering design. This omission is not intended to minimize the importance of trench construction and the potential for failure under these circumstances. However trench sheeting and bracing often are selected by a contractor based on experience without the participation of a professional civil engineer. For such a situation there are various standard rules in building codes or in Federal safety and health regulations. Such empirical rules are not included or utilized herein.

1.2 Characteristics of the Supported Excavation

The essential elements of an excavation support system comprise a wall, supports, and a reaction to or anchorage for the loads transmitted through the bracing. This

1

apparently simple arrangement has a number of features which distinguish it from ordinary civil engineering structures:

1. Installation of the support system is coordinated with the excavation and is generally built from the top downward. Excavation and support installation usually are performed in stages and the critical condition for various elements of the support system may occur at some intermediate stage during this construction.

2. The support system usually is temporary in nature with a life ranging from several months to several years. In special cases the wall of the excavation is incorporated into the permanent structure.

3. In some cases safety factors are lower and the stress levels higher than for a permanent structure. The choice of safety factor is influenced by the consequences of failure as they affect the work within the project and adjacent properties.

4. Deformation of the support system can be as important as its load-carrying capacity. Numerous instances have occurred where the braced excavation is safe against an overall failure but movements of the retained soil are unacceptable because of their effects on interior or exterior structures.

1.3 Difficulties with Supported Excavations

Braced and tied-back excavations may lead to difficulties. These problems usually fall in one of the three following groups.

1. Structural inadequacy of the support system which arises from one of the following:

   a. Errors in selection of earth or water pressures, including failure to consider special loadings such as heavy surcharges, an earth slope above the top of the support system, or insufficient passive resistance of interior berms.

   b. Oversights in structural design, often involving weakness of connection details and failure to take into account practical problems of installation.

   c. Inappropriate construction procedures, particularly careless installation of the support wall or failure to match elements of the bracing system.

d.   Unforeseen adverse subsurface conditions.

e.   Poor coordination with other construction activities in the interior of the excavation.

2.   Excessive movement or deflection of the system may be a product of the structural weaknesses described above or one of the following:

a.   Insufficient stiffness of the support system, including lack of fit of connections.

b.   Inattention to support requirements during construction including failure to pre-load braces, over-deepening of excavation beneath support levels, delays in adding or supplementing support and premature removal of support.

c.   Loss of ground from over-excavation made for placement of support wall or lagging, excessive lengths of slurry wall panels, or inadequate backfill of holes pre-drilled for piles.

d.   Ineffective ground water control, resulting in piping of soil particles through openings in the support wall or "quick" conditions at the excavated subgrade.

e.   The presence of deep soft clay soils beneath or in the lower portion of the excavation.

3.   Movements of the surrounding ground caused by excavation and construction of the support system, but not related to the support system itself, can occur as follows:

a.   Settlement due to the increase in soil effective stresses which accompanies drawdown.

b.   Erosion or piping due to surface water runoff, ground water gradients, or utilities leaking into the excavation.

c.   Settlement caused by pile driving, blasting, or mechanical vibrations in the construction.

d.   Weathering or deterioration of soil and rock exposed in the cut face or at subgrade.

Many of these factors are interrelated. For example driving of steel sheet piling might settle loose sand retained by the bracing system. This settlement in turn

3

could break a water main in the retained sand. The consequent increase in water pressures or piping could damage the support system that has been selected under the assumption of a dewatered condition.

There are three spheres of exposure to damage: the working space within the excavation; areas immediately outside the excavation; and surrounding areas at some distance from the excavation. Within the excavation these hazards threaten those most directly involved in the project and ordinarily are the first concern of the designer. Within adjacent exterior areas there is the risk of injury to third parties plus damage to structures founded on or within soil and rock which are retained by the excavation support system. Their protection influences strongly the requirements for the excavation support scheme. Within the third sphere the risk is less but damage and nuisance may still occur. The ultimate impact can be serious because of the larger area involved and the likelihood that effects are unforeseen.

1.4   Assignment of Responsibility

Opinions diverge among engineers, constructors and public officials regarding assignment of responsibility for performance of an excavation support system. This is reflected in the following three general procedures for design and construction:

1.   Design of the support system is by an agent of the owner of the permanent structure, generally by the structural engineer for the project, and it is executed by a contractor selected by competitive bidding or negotiation. Inspection of the work is by an agent of the owner.

2.   Design and construction are by a contractor selected by the owner, either the builder of the permanent structure or a sub-contractor specializing in excavation and bracing work. Owner's agents provide only periodic inspection supplementing that of local building officials.

3.   The excavation and support system are undertaken according to empirical rules embodied in building codes regulating such work and a professional engineer may not be involved. Inspection, approval or review is by representatives of the agency that administers the code.

Each of these procedures has advantages appropriate to certain building situations and each has it particular problems. In many projects elements of all three schemes are intermingled. In any case, results depend on the skill and integrity of those involved in the design, construction and in administration of the building code. The first procedure tends to keep the interest and liability of the owner under the owner's direct control. The second relies on the skill of a specialist contracting organization throughout. For major work less reliance is being placed on code provisions, empirical design, and inspection by public agencies, as embodied in the third procedure.

1.5   Definitions of Parties of the Work

The parties involved in the work of braced excavation are referred to in subsequent chapters of the manual and are defined as follows:

Owner is the party sponsoring the new permanent construction which requires a braced or tied-back excavation. The owner engages design services in connection with the work and contracts with a builder for construction of the work.

Engineer-of-Record ordinarily is the professional civil engineer retained by the owner to design the permanent construction. Depending on local laws and the engineer's arrangement with the owner, the engineer may also be responsible for some elements of the excavation and support system design. The engineer's professional stamp appears on contract drawings relating to the excavation. The engineer stipulates the general requirements of the excavation and assesses its compliance with applicable codes. The engineer will also have a part in reviewing the contractor's shop drawings as the owner's representative.

Designer is the engineer, presumably a professional civil engineer, who is directly involved in technical decisions for the excavation. The designer may be working with the contractor or the engineer-of-record and is concerned ordinarily with technical aspects and not contractual or administrative questions.

Resident Engineer will direct the field inspection of a braced excavation. The resident engineer may be an employee of the owner or the engineer-of-record. The resident engineer is to see that specification provisions are followed and will be concerned with departures from the contract, field expedients, and alternative procedures. Ultimately, the resident engineer will share with the contractor much of the responsibility for field decisions. The resident engineer's staff includes inspectors whose work the resident engineer must oversee, who are technically qualified but are not necessarily professional engineers.

Geotechnical Engineer has evaluated site and subsurface conditions in advance of construction, has made recommendations regarding the permanent structure and has given criteria relating to the excavation and

5

support system. The geotechnical engineer would have been retained by the owner, the engineer-of-record or, less frequently, by the contractor.

Contractor has a contract with the owner for construction of the excavation support system. This may be the prime contractor or it may refer to specialist subcontractors involved in particular elements of the work, such as construction of the enclosing wall, interior excavation, underpinning, or tieback installations. The contractor may employ or retain the engineer who designs the excavation support system.

1.6   Objectives and Organization of the Guidelines

These guidelines do not advocate any single approach to carrying out the work of braced and tied-back excavations or the related assignment of responsibility. These guidelines describe various techniques used to support excavations, methods for design and potential difficulties. Recommendations are provided for dealing with problems or for anticipating them in specifications. Criteria that should be considered in a good design are outlined and conditions under which these criteria should be modified are pointed out. Ultimately the owner, engineer-of-record, specialized engineering consultants, contractor and the public building official all contribute to the safety and quality of the support system.

While there is no intent that the guidelines play a part in litigation, they might conceivably be utilized to challenge engineers' decisions in cases where damage has been suffered. *In this respect it is stated categorically that decisions of a professional civil engineer directly involved in the specific case should reasonably be assumed to take precedence over generalizations contained in these guidelines.*

Safety of the support system and the protection of facilities within and adjoining the excavation are a primary concern of these guidelines. However, the safety work rules and requirements that are within the purview of the Occupational Safety and Health Administration (OSHA) generally are not treated herein.

1.6.1 Readership. The guidelines are directed principally to civil engineers without distinction as to whether they are associated with the owner, design organization, contractor or public building agency. It is assumed that the engineers will have experience and competence in design, but may not possess specialized knowledge in geotechnical engineering. The guidelines are not intended to be a handbook for design to be used by those who have little knowledge of construction, structural design or geotechnical engineering. Although not a substitute for a building code, it is hoped that professional civil engineers on the staff of public agencies enforcing code provisions will find it useful. It is intended that the manual reflects what a group of specialists in this subject consider to be good engineering practice.

1.6.2 Organization of Contents. The main body of the manual consists of eight chapters. No attempt is made to present detailed technical information readily available in standard sources. References that are necessary to implement recommendations of the guidelines or which amplify important topics are listed at the end of each chapter.

Chapters 2, 3 and 4 treat background subjects: investigations, water control, design loadings, and stability considerations. These chapters address work which chiefly concerns the geotechnical engineer. Chapters 5, 6 and 7, on walls, internal support systems and tiebacks, respectively, relate primarily to the work of the designer. Chapter 8 deals with problems of installation, overall system stability and control of construction which are the concerns of the engineer-of-record, the resident engineer, and the contractor.

CHAPTER 2:
INVESTIGATIONS FOR DESIGN AND FOR CONSTRUCTION CONTROL

2.1 Introduction

This chapter addresses features of a geotechnical investigation program which focus on braced and tied-back excavations. Such features would ordinarily be part of a wider geotechnical study for the entire project. Presumably general considerations have established the need for an excavation support system. An investigation for design and construction control involves four aspects:

1. Subsurface exploration employing borings and test pits to define soil and rock strata and ground water conditions.

2. Laboratory and field testing to establish strength, compressibility and permeability of subsurface materials for design of the support system.

3. Preconstruction surveys to establish the condition of adjacent properties and utilities.

4. Monitoring during construction of: ground movements, piezometric levels, performance of the support system and effects of excavation on adjacent facilities.

More than other facets of these guidelines, the investigation activity involves areas of geotechnical engineering that are controlled by standards and references. These include the following subject areas:

1. ASTM standard procedures for boring, sampling, field and laboratory testing are outlined in Reference 2-1. In the absence of building code requirements or special procedures dictated by the engineer-of-record or the geotechnical engineer, these standards should generally prevail.

2. Definitions of geotechnical terms and soil and rock materials utilized in these guidelines conform to the "Glossary of Terms and Definitions in Soil Mechanics" (Reference 2-2). This follows the Unified Soil Classification System which has been adopted nationwide by Federal agencies. Investigations covered by these guidelines should utilize that classification system.

3. Planning of subsurface investigations should consider the ASCE Manual of Practice, "Subsurface Investigation for Design and Construction of Foundations of Buildings" (Reference 2-3).

8

4. Procedures for field determination of properties are reviewed in various papers and proceedings, notably References 2-4 through 2-7.

5. Devices for field measurement of movement and deformation are reviewed in References 2-8 through 2-11.

2.2 Subsurface Investigation

The scope of an investigation for a particular site depends on the scale of the excavation, nature of subsurface materials, and need for protection of adjacent properties. For major work, the investigation might consist of a preliminary phase followed by a final design phase. The preliminary phase includes a limited number of fairly widely-spaced borings to define overall subsurface conditions and to identify problem areas. The standard split-spoon is used frequently in this phase to obtain soil samples for identification. Based on the results of the preliminary study, a more detailed boring program might prove necessary for final design, including undisturbed samples for laboratory tests and observation wells at key locations.

2.2.1 **Special Requirements Relating to Braced Excavations.** The number of borings in the complete program may be controlled by code provisions or considerations of the permanent structure. However, to evaluate excavation support requirements the following factors should be considered:

1. Special ground water conditions such as the presence of artesian pressure or perched water can have a critical effect on construction. Experience with soil running up into a borehole during withdrawal of drilling tools, loss of drilling fluid, or sloughing of soil from walls of the borehole should be carefully recorded.

2. In boring operations any evidence of extremely compact material, cobbles, boulders, cemented layers or lenses, "hardpan", "ironpan", rubble fill, debris or other material that would impede the installation of a an excavation wall should be noted. It is absolutely essential that any such evidence be recorded on boring logs to be included in contract information for bidders.

3. At a site where rock or rock-like materials are present within the excavation, the borings must define the character of materials where there is a transition between residual soil and bedrock, and should locate the surface of materials of rock-like hardness.

4. The boring procedures should facilitate the making of a continuous record of the character of materials encountered. Modern heavy-duty rotary

equipment such as truck-mounted rotary drills operating grinding bits in mudded boreholes, or continuous large-diameter hollow-stem augers can obscure ground water levels, do not provide a record of casing blows, and can pass through the upper decomposed bedrock zone as if it were merely compact soil. Care should be taken that this ease of penetration does not prove misleading as to the condition of the in-situ materials.

2.2.2 Boring and Sampling Procedures. Basic techniques for boring and sampling are defined by the ASTM Standards of Reference 2-1. Soil samples often are obtained using a standard two-inch (50mm) split-spoon, thin-walled "Shelby" tube samplers typically two to three inches (50-75mm) in diameter, or thin-walled "undisturbed", fixed-piston samplers which are three inches (75mm) or larger in diameter. Undisturbed sampling equipment conforms to certain technical requirements outlined in Reference 2-7. Split-spoon samples are used for visual identification and for laboratory classification tests. Shelby tube samples provide evidence of soil structure with less disturbance than spoon samples, permit more detailed identification tests, and may be useful for simple unconfined compression tests. Undisturbed samples are utilized for tests of engineering properties including determination of consolidation and strength characteristics.

Many project investigations emphasize split-spoon sampling and simple visual classification of samples with minimum laboratory testing. Subsoil characteristics and design parameters are then estimated from correlations with standard penetration resistance and from previous local experience and building code provisions. It should be recognized that much of the subsequent discussion in this chapter involves techniques appropriate for projects which are larger or more elaborate than the ordinary.

2.2.3 Rock Core Drilling and Logging. In major investigations, diamond-drill coring is performed with a double-tube core barrel yielding a core of NX-size, approximately 2-1/8 inches (50mm) in diameter. Where bedrock lies below the base of excavation and rock characteristics are of secondary interest, smaller diameter core may be acceptable. Where rock conditions are of prime importance it may be desirable to use the triple-tube core barrel wherein the core is surrounded by a thin liner split longitudinally inside a recovery barrel within the coring barrel. This permits the core in its liner to be placed intact in the core box without being dumped from the tube. Logging of the core should include the following information as a minimum:

1. General description of rock type, mineral fabric, foliation or lineation, and degree of weathering.

2. Presence of discontinuities, that is: joints, fractures, faults, and shears, with particular attention to zones of possible movement. The dip and spacing of discontinuities should be noted with their relation to the overall foliation.

Joints should be described with respect to smoothness and filling or coating material.

3. Percentage of core recovery and rock quality designation (RQD) (Reference 2-12). Care must be taken that percentage recovery is not overestimated but is based on the core reconstituted in its densest possible packing, rather than laid loosely in the core box.

2.2.4 Special Investigation Techniques. Apart from test borings, certain special investigation methods should be given consideration for deep braced excavations. The following techniques may be useful where the nature of subsurface materials makes it difficult to obtain satisfactory samples for laboratory testing or where information is needed on in-situ characteristics.

1. Deep Test Pits. Hand-dug and braced test pits can reveal conditions not exposed by borings. These include the difficulty of installing lagging against certain soils, the prospects of inflow from perched water, difficulties of digging including the effects of obstructions, hardness of residual soils or decomposed rock and the depth at which practical rock excavation commences. It also permits taking of hand-cut samples suitable for detailed laboratory tests. Exploratory pits are often necessary to determine the character of buried portions of adjacent structures. A complete and detailed record must be made of digging conditions encountered in addition to the ordinary description of materials. While hand-dug pits are particularly useful, a more common and less costly procedure is the machine excavation of inspection pits by large-diameter auger, caisson rigs or backhoes.

2. In-Situ Strength Testing. Equipment for in-situ strength testing includes cone penetrometers, vane shear devices, pressuremeters and other equipment as described in References 2-4 through 2-6. Use of these devices can provide index values or strength parameters that will increase the understanding of conditions to be encountered in excavations. For example, the vane shear test is particularly useful to define the trend of the undrained shear strength profile with depth and to identify particularly sensitive soils. A cone device can be used to help quantify shear strength and develop the soil profile. The pressuremeter is applicable to assessment of soil strength, modulus and lateral stress levels if the test is performed in carefully prepared holes or if a "self-boring" pressuremeter is used.

3. Determination of In-Situ Rock Characteristics. Special devices for observing rock in-situ include the borehole camera which permits viewing of rock strata on a screen or by means of a photographic record and various core orientation methods. Procedures are available for estimating residual stresses or rock strength and stress-strain properties in-situ as described in References 2-8, 2-13, and 2-14. These techniques usually are reserved for

11

deep rock excavation or mined openings in rock. For an excavation which bottoms below the bedrock surface it can be particularly important to determine the attitude and condition of joints and faults that will affect stability of the rock face. For this purpose, various means of orienting the core barrel are available, but the most reliable information comes from the performance of the local rock in previous excavations. Full advantage should be taken of examination of outcrops and exposures to assess bedrock structural characteristics.

4.   Field Permeability and Pumping Tests. Permeability tests may range from simple falling-head tests made in boreholes to full-scale deep-well pumping tests performed with an array of observation wells to monitor drawdown. In the minimum test boring program, information on field permeability can be obtained by means of rising or falling head tests. Generally it is only for more elaborate large-scale excavations that well pumping tests are justified. However, if it is essential that information be gathered on dewatering problems in the excavation, a pumping test is the most reliable means of modeling the prototype operation on a near full-scale.

2.3   Soil Laboratory Testing

Laboratory tests are employed for braced excavations chiefly to delineate subsoil strata and to determine strength and stress-strain properties. Identification procedures include determination of moisture content, unit weight, Atterberg plasticity limits, and grain size characteristics and can be performed on disturbed samples which contain all constituent particles. Identification tests should provide a proper classification of soils and establish fundamental properties for designers and bidders. Water content and Atterberg limit tests on clay or silt samples help to clarify both preload history and overall consistency of the material. Grain size determinations are useful in relation to dewatering problems and the stability of materials to be excavated beneath the water table.

2.3.1 Tests for Shear Strength. A discussion of strength parameters to be utilized in evaluating loads on an excavation support system and overall stability is presented in Section 4.9. Triaxial strength testing should be reserved for "undisturbed" samples with diameters as sampled not less than about three inches (75mm). Unconfined compression testing of two-inch (50mm) Shelby tube samples is essentially an identification procedure yielding an index property chiefly useful for correlation purposes. While analysis of bearing capacity of the permanent structure may dictate other, more elaborate tests of strength properties, the braced excavation design usually involves the following types of tests:

1.   For construction conditions where clays or impervious clay-silt-sand mixtures undergo little volume change during excavation, unconsolidated-

12

undrained triaxial compression tests are chiefly applicable. These yield a specific value of undrained shear strength with friction angle assumed as zero. Testing should establish a profile of strength versus depth which must be consistent with the estimated preconsolidation condition that controls undrained strengths. Testing should also define the soil strength in the disturbed state so as to identify sensitivity. A highly sensitive clay presents a more difficult design condition than a non-sensitive one.

2.   In overseas practice, significant use has been made of in-situ vane shear tests to evaluate the undrained shear strength of clays at deep excavation sites (e. g., References 2-15 and 2-16). Vane shear testing can be rapid, relatively economical, and reproducible in homogeneous deposits of soft to medium clay. Vane shear tests provide a measure of peak and residual shear strengths. At a particular site, vane tests in soft and sensitive clays often show less scatter than the shear strength results from unconfined compression, pressuremeter, and other testing techniques. Because the theoretical nature of the vane failure mechanism is not fully understood, caution must be exercised in interpreting vane data. Correction factors have been proposed to correlate measured vane strengths with back-calculated strengths from the analyses of field failures (References 2-17 and 2-18). Recent study has shown that correction factors correlate with the ratio of vane shear strength to effective vertical overburden stress, and that this undrained strength ratio can be estimated on the basis of plasticity index and degree of overconsolidation (Reference 2-19).

3.   In special cases involving deep deposits of soft clay it is advisable to consider performing a supplemental series of triaxial extension and direct simple shear tests. Such a procedure simulates the loading path exerted on the soil. Experience has shown that strength in the passive and radial shear zones can be lower than that in the active zone due to soil anisotropy. These effects lead to a lower stability condition than might be assumed on the basis of compression tests only (Reference 2-20).

4.   For "cohesionless" soils, that is, relatively clean coarse-grained materials, and for stiff, moderately to heavily over-consolidated clays, drained strength parameters are generally appropriate for evaluation of end-of-construction conditions. Moderately to heavily overconsolidated clays develop negative pore-water pressures when subjected to shear stresses induced by excavation. These pressures dissipate with time, causing a decrease in shear strength consistent with drained strength parameters.

Drained strengths may be determined either from fully drained slow tests or from consolidated-undrained triaxial tests with pore pressure measurements. For some sands it may be more expedient to obtain drained strength parameters by direct shear procedures. In any case, the shearing force should be applied slowly enough so that full volume change can take place and no excess pore pressure develops during shear. As a practical matter, it should be recognized that the shear strength of coarse-grained materials is much more commonly estimated from in-situ penetration tests than by laboratory testing.

5. A difficult analytic problem occurs in the case of very heavily overconsolidated fissured or slickensided plastic clays. In this case, undrained shear tests can yield extremely erratic strengths with average values so high that they are inappropriate for stability analysis or for design of the bracing system. As an alternative, it might be advisable to perform shear tests in a fully drained condition yielding an effective angle of drained shearing resistance. Often, as an expedient, local practice has established a reasonable set of design parameters.

6. In special cases where extremely soft and essentially normally consolidated fine-grained soil is to be retained by a braced excavation, a situation may arise through loading or drawdown in which shear strength will increase significantly during construction. In this case consolidated-undrained tests may be appropriate to estimate an increase in strength with the passage of time.

7. It is essential that the test conditions under which strength parameters are determined are clearly recorded for the benefit of the designer and bidders. Strength parameters should be described in terms of total or effective stress and the geotechnical investigators should make a logical correlation of these properties with subsoil stratification obtained from borings.

2.3.2 Consolidation Testing. Consolidation testing usually is of interest in investigations for the permanent structure and in correlating shear strength with pre-stress conditions. However, it may also be appropriate in analyzing settlements produced by construction drawdown or to provide the basis for analysis of deformations of the surrounding ground by finite element procedures. If consolidation tests are necessary the following factors should be considered:

1. No theoretical analysis of the subtle settlement effects of drawdown can be undertaken without first making a plausible determination of the profile of preconsolidation stress in the surrounding soil. This can only be attempted if high quality undisturbed samples have been obtained and if some of these samples are at least moderately plastic so as to yield a well-defined preconsolidation value in the laboratory pressure-void-ratio curve.

2. At least some of the consolidation tests should include an intermediate cycle of rebound and reloading to delineate accurately compression characteristics in the range below the preconsolidation stress. A settlement analysis should

14

not be made utilizing the laboratory pressure-void-ratio curve in an initial cycle of loading at stresses lower than the preconsolidation stress. A detailed description of procedures for testing and evaluation of preconsolidation conditions and recompression characteristics is given in Reference 2-21.

The most reliable source of information on the effect of drawdown is found in experiences with earlier deep excavations nearby. Use of such records requires sufficient information on subsoils and drawdown to permit a valid comparison with the new project plan.

2.3.3 Stress-Strain Properties. Stress-strain properties should not be computed from an initial cycle of loading in a triaxial shear (or a consolidation test). At least one unload-reload cycle of loading should be performed. Loading paths which model those anticipated during excavation should also be considered.

2.3.4 Sensitivity to Vibration. Loose cohesionless soils in the vicinity of the braced excavation may be subject to densification and consequent settlement due to vibrations from driving of piles and sheeting or from blasting for the excavation. The degree of vulnerability is difficult to assess but is indicated by the in-situ relative density estimated from standard sampler penetration resistance or by direct measurement of undisturbed sample densities. As with other problems of soil deformation, the most reliable evaluation is obtained from case histories of excavation in nearby sites or in similar ground conditions.

2.3.5 Permeability Testing. Laboratory permeability testing in connection with a braced excavation is of limited usefulness. Testing of fairly clean sands that lack lenses or seams of silty material can supplement field tests of permeability in-situ. However, if problems of ground water control are of particular importance, emphasis should be placed on field exploration and in-situ testing procedures to determine permeability.

2.4   Preconstruction Survey of Adjacent Properties and Utilities

Surveys are made before the start of excavation, pile driving, and dewatering to ascertain conditions of adjacent properties and utilities prior to any possible disturbance during construction. The importance of the investigation will depend on the degree of development of surrounding areas, size of the excavation and potential threat of disturbance posed by the new work. A systematic preconstruction survey is desirable whenever third party interests are involved. The following factors should be considered in planning and implementing the survey:

1. The construction specifications should assign tasks and stipulate the division of responsibility to fill any gaps in the information that may have been collected by the engineer-of-record during design. For actual examination of structures it is desirable to have a joint party consisting of engineer's and contractor's representatives making a photographic record and a tape-recorded description of conditions. When completed, this report should be attested by field party members and submitted to the building owner for review and, if possible, assent as to its completeness and accuracy.

2. Inspection of existing structures should concentrate on evidence of prior damage, cracking, distortion, weaknesses in structural elements, deterioration, corrosion, excessive stress, overloading, or use of the structure in a manner which may not have been intended in its design. At this stage it may be desirable to install telltale marks or measuring points at key locations on the structure which will be used to assess movements.

3. If a question arises concerning the need for underpinning or protection of the existing structure, a search must be made for plans showing as-built foundations and a decision made whether these need to be investigated by test pits, probings or core holes through foundation elements. Necessary investigations of existing foundations should be done before construction so that the designer and bidders can better evaluate strategies for protection. It is of vital importance that any records of existing foundations obtained or developed in investigations be assembled in orderly fashion and their presence made known to bidders.

4. Often a reasonably complete picture of adjacent utilities or utilities entering the site of the planned excavation will be assembled by the engineer-of-record for use in the engineer's design. Utility lines which will be intercepted, blocked or relocated must be described in the greatest detail possible for the construction contract. Specifications should call attention to known or suspected locations of these utilities and procedures required for their treatment. Availability of this information does not relieve the contractor from making an independent evaluation of utilities that might be affected by the particular scheme being considered by the contractor for the supported excavation. While the contractor cannot be expected to make a detailed utility survey in the bidding period, the successful bidder must make a detailed survey of any additional utilities whose disposition is not provided for in the specifications.

5. Possibly no element of the investigation relating to braced excavations is so important as collecting, verifying, and documenting data on the condition of utilities that will be intercepted or affected by the excavation. Existing conditions may have to be clarified by exploratory pits, probing or

geophysical surveys. It is essential that all information developed by the designer be assembled in an orderly fashion for examination by the bidders.

2.5   Field Observations During Construction

Field observations and instrumentation are used to monitor movements of adjacent ground and structures, load and strain in the excavation support system, and piezometric levels inside and outside the excavation. Many types of measuring devices and observation techniques are now available which are discussed briefly in the following section. The role of measurements in construction control is treated in Chapter 8. Specifications should define the division of responsibility for field observations. Routine survey measurements including those that can be made by optical level or by direct steel taping usually are performed by the contractor's field engineering staff. More elaborate observations utilizing special devices might be carried out by the resident engineer's representatives. Procedures for timely and systematic processing of the data are essential so that results may be available to the resident engineer and the contractor's superintendent.

2.5.1 Surface Measuring Points. Settlement points, telltale markers or scribed lines can be established on buildings, pavement or other permanent features to monitor movement during construction. Generally the simplest and most obvious measurements are also the most useful. Periodic readings are taken during construction to determine changes in elevation or dimension. Under favorable weather and site conditions and with unmoving benchmarks available, an accuracy in the range of 0.005 to 0.01 feet (1/16 to 1/8 inch, or 2 to 3mm is practicable in an optical level survey where error is balanced in a circuit closing back on a bench. Visual evidence of cracking or the use of tell-tales on plastered walls are useful, but quantitative data should also be obtained by taping horizontal distances between points established on lines perpendicular to the excavation wall. For accurate settlement observations, particularly under conditions where the excavation produces small but widespread movements of the surrounding ground or where construction dewatering is extensive, a deep stable benchmark may be needed. These can consist of fixed reference points established on buildings with deep foundations or which are provided by rods in cased boreholes whose points are driven into an unyielding stratum at a depth below the zone of movement.

2.5.2 Inclinometer Devices. Inclinometers are useful in determining a profile of lateral movements of wall components or of soils on a vertical line outside the excavation. The unit is installed by anchoring a grooved casing in a borehole so that measurements of slope from the vertical can be made in a known compass direction utilizing a pendulum or accelerometer device lowered along the casing. Modern equipment with experienced observers can yield an accuracy of horizontal movements equal to about 1/4 inch (6mm) in a 100-foot (30m) depth. This degree of accuracy generally suffices for practical control. They can be of great value in assessing the mode and progression of ground movements that affect adjacent structures. Use of this device is a specialized operation which requires carefully framed specifications, employment of specialist subcontractors for installation and measurements by skilled and experienced personnel. Inclinometers should not be utilized unless these requirements are satisfied.

2.5.3 Measurements on Excavation Support System. These consist of five principal categories of equipment:

1. Strain gauges attached to main bracing members or load cells to measure elastic axial compression of braces or loads in ties due to the application of external lateral pressures. Alternatively with braces, jacks and jacking plates can be used to lift the brace off of the wall to directly assess the load. Caution should be exercised when evaluating brace loads by hydraulic jacking to ensure that calibrated jacks are used. Experience has shown that ram friction, eccentric loading, and improper calibration contribute to inaccuracies such that hydraulic jacks commonly overestimate load. Independent confirmation by a load cell is recommended.

2. Accurate gauge points placed for a similar purpose on bracing members where extensometer dial gauges measure the strain.

3. Accurate measurements of dimensional changes between fixed elements on opposite sides of the excavation.

4. Extensometers incorporating either single or multiple position readings are installed through the excavation wall into the retained soil or rock to measure movements with respect to an anchor point.

5. Vertical extensometers are installed inside the excavation to measure heave and outside the excavation to measure settlement. Extensometers permit the monitoring of vertical displacements at various depths. Such measurements can be a useful supplement to surface and building surveys surrounding the excavation, and may provide critical information about plastic deformation at the bottom of the excavation under circumstances where base movements and failure are important concerns.

Each of these schemes involves various specialized techniques in installation and observation. Some difficulties are discussed in Chapter 8. Ordinarily, these procedures should not be undertaken without the collaboration of specialists either in the employ of the engineer-of-record or of the contractor. Wasted effort and misleading information may result unless a program is planned specifically for the practical purposes of construction control.

2.6   Availability of Records of Field Investigations

Investigations carried out in the design stage before bidding must be completely and systematically documented in such a manner that the record can be used for the designer and bidders and for construction control.   Data from physical investigations, such as test borings or test pits, the preconstruction survey, and records of utilities should be assembled in some conventional and orderly fashion. Although these data are not necessarily included within the actual body of the contract documents, each item should be described as information available to bidders.   Data should be placed in a form such that it can be readily referred to by bidders and furnished to them in such a manner that bidders can master the essential features without an unreasonably long and detailed study requiring the assistance of specialist consultants for a protracted time.

2.7   References

2-1   ASTM (1997). *Annual Book of ASTM Standards, Soil and Rock.* American Society for Testing Materials, Philadelphia, PA.

2-2   Committee of the Soil Mechanics and Foundations Division on Glossary of Terms and Definitions and on Soil Classification (1941). "Soil Mechanics Nomenclature," *ASCE Manual of Engineering Practice No. 22,* ASCE, New York, NY.

2-3   Committee for the Manual on Subsurface Investigations (1976). "Subsurface Investigation for Design and Construction of Foundations of Buildings," *ASCE Manuals and Reports on Engineering Practice No. 56,* ASCE, New York, NY.

2-4   Clemence, S. P., Editor (1986). "Use of In-Situ Tests in Geotechnical Engineering," *Geotechnical Special Publication No. 6,* ASCE, New York, NY 1986.

2-5   Jamiolkowski, B. M., Ladd, C. C., Germaine, J. T., and Lancellotta, R. (1985). "New Developments in Field and Laboratory Testing of Soils," *Proceedings,* 11th International Conference on Soil Mechanics and Foundation Engineering, San Francisco, CA, Theme Lecture, Session 2.

2-6   ASCE (1975). *Proceedings of the Conference on In-Situ Measurements of Soil Properties,* Geotechnical Engineering Division, ASCE, Raleigh, NC.

2-7   Hvorslev, M. J. (1948). *Subsurface Exploration and Sampling of Soils for Civil Engineering Purposes*, U. S. Army Engineer Waterways Experiment Station, Vicksburg, MS.

2-8   Dunnicliff, J. (1988). *Geotechnical Instrumentation for Monitoring Field Performance*, John Wiley and Sons, New York, NY.

2-9   Cording, E. J., Henderson, A. J., MacPherson, H. H., Hansmire, W. H., Jones, R. A., Mahar, J. W., and O'Rourke, T. D. (1975). "Methods for Geotechnical Observations and Instrumentation in Tunneling," *Report No. UILU-ENG 75 2022, Vols. 1 and 2*, National Science Foundation, Washington, D. C.

2-10   Thompson, D. E., Edgers, L., Mooney, J. S., Young, L. W., and Wall, C. F. (1983). "Field Evaluation of Advanced Methods of Geotechnical Instrumentation for Transit Tunneling," *Report No. UMTA-MA-06-0100-83-2*, U.S. Department of Transportation, Washington, D. C.

2-11   Gould, J. P. and Dunnicliff, J. (1971). "Accuracy of Field Deformation Measurements," *Proceedings*, 4th Pan American Conference on Soil Mechanics and Foundation Engineering, San Juan, PR, ASCE, New York, NY, Vol. 1, 313-366.

2-12   Deere, D. U. and Deere, D. W. (1989). "Rock Quality Designation (RQD) After Twenty Years," *Contract Report GL-89-1*, Department of the Army, Washington, D. C., 100 p.

2-13   Franklin, J. A., and Dusseault, M. B. (1989). *Rock Engineering*, McGraw-Hill Publishing Co., New York, NY.

2-14   Brady, B. H. G., and Brown, E. T. (1985). *Rock Mechanics for Underground Mining*, George Allen & Unwin, Boston, MA.

2-15   Bjerrum, L. and Eide, O. (1956). "Stability of Strutted Excavations in Clay," *Geotechnique*, 6(1), 32-47.

2-16   Karlsrud, K. (1983). "Performance and Design of Slurry Walls in Soft Clay," *Publication Report No. 149*, Norwegian Geotechnical Institute, Oslo.

2-17   Bjerrum, L. (1972). "Embankments on Soft Ground," *Proceedings*, ASCE Specialty Conference on Performance of Earth and Earth-supported Structures, Lafayette, IN, Vol. 2, 1-54.

2-18   Mesri, G. (1975). Discussion of "New Design Procedure for Stability of Soft Clays," by C. C. Ladd and R. Foott. *Journal of the Geotechnical Engineering Division*, ASCE, 101(GT4), 409-412.

2-19   Aas, G., Lacasse, S., Lunne, T., and Hoeg, K. (1986). "Use of In-Situ Tests for Foundation Design on Clay," *Use of In-Situ Tests in Geotechnical Engineering (GSP 6)*, Ed. S. P. Clemence, ASCE, New York, NY, 1-30.

2-20   Ladd, C. C. and Foott, R. (1974). "New Design Procedures for Stability of Soft Clays," *Journal of the Geotechnical Engineering Division*, ASCE, 100(GT7), 763-786.

2-21   Brumond, W. F., Jonas, E., and Ladd, C. C. (1976). "Estimation of Consolidation Settlement," *Manual of Practice Special Report 163*, Transportation Research Board, Washington, D. C.

CHAPTER 3: CONTROL OF SURFACE WATER AND GROUNDWATER

3.1   Need for Water Control

Braced excavations require control of groundwater when it adversely affects construction or performance of the support system. The need for and the impact of a water control system should be studied before construction. Its purposes are:

1.   To permit excavation in the dry and to reduce water pressures acting on the support system.

2.   To minimize disturbance, heave, or softening of the excavated bottom and to prevent piping or flow of materials through the wall elements.

3.   To increase passive resistance of interior berms (or of the interior subgrade) to lateral inward movement.

4.   To aid construction of components, such as tiebacks, reaction elements for raking braces, elevator pits, permanent under-drainage systems and the like.

Information useful for the analysis of various dewatering systems is contained in References 3-1 through 3-4. Chapter 4 considers the effect of water pressures on the retaining structure loading. Procedures and equipment for water control are continually evolving and are described in technical periodicals and trade manuals. No attempt is made to present such information herein.

3.2   Division of Tasks

Design of the water control system normally is viewed as a construction expedient and, traditionally, has fallen within the contractor's responsibility. However, water control activities can have serious impact on adjacent properties and on the permanent facility to be constructed within the excavation. Consequently, both the owner's engineer and the contractor's engineer are likely to become involved in the functioning of a groundwater control system. Tasks of the owner's engineer include some or all of the following:

1.   Investigate and present technical data relating to existing groundwater conditions.

2.   Assess the need for groundwater control in making the excavation, its effect on adjacent structures, and the necessity for recharge of the exterior water table.

22

3. Formulate dewatering specifications, including provisions for construction monitoring and requirements for and limitations on drawdown.

4. Review the contractor's proposed dewatering methods for compliance with specifications.

Tasks of the contractor's engineer include some or all of the following:

1. Plan a dewatering system to meet specifications and the job needs. Frequently this planning is aided by a specialist subcontractor.

2. Assess the influence of the contractor's method of construction on the water control scheme.

3. Evaluate the protection afforded the permanent structure and effects on existing adjacent structures.

Ultimately, when the system and installation method are the contractor's choice, responsibility for adequacy of the system rests with the contractor.

3.3   Field Investigations

In addition to the determination of subsurface conditions discussed in Chapter 2, pre-construction exploration must take into account the data needed to plan a water control program. Included in a "routine investigation" should be the measurement of groundwater levels within and around the intended excavation to a level at least 1.5 times the maximum depth of the excavation below the highest water surface observed in the borings. No test boring records should be accepted without a statement of the water levels observed at each borehole or an explanation for the absence of such information or its lack of reliability.

The exploration program generally will be in one of the following three classes:

1. A minimum investigation includes: determination of groundwater or piezometric levels in each boring at the time of investigation; noting the presence of perched, depressed, or artesian water levels; and describing any occurrence of running soil or loss of wash water. Also, grain size analyses of potential aquifer soils should be performed. The D_{10} value so determined can be used through empirical correlations to estimate permeability (Reference 3-1). When using grain size data to estimate permeability, it should be recognized that this approach may discount the effects of in-situ structure. For example, thin discrete layers of sand can result in a relatively high horizontal permeability, even though a significant fraction of fines are indicated by the grain size analysis.

2. Detailed investigations include in addition: in-situ permeability determinations by falling or rising head tests in borings; installation of observation wells or simple piezometers to measure water levels, continuing observations over a sufficient time period to establish the typical variation in levels; and research of local references and case histories to assess water table fluctuations.

3. Special investigations include in addition: full-scale deep well pumping tests or water pressure tests between packers in cased boreholes; study of the original and present conditions of nearby water-carrying utilities; undisturbed sampling for laboratory permeability testing; and installation of sensitive piezometers for observation of water table fluctuations. In some cases the groundwater is sampled for tests of chemical constituents, dissolved or suspended materials or potential corrosion characteristics.

Exploration information and a study of geological and geotechnical references should provide answers to the following key questions:

1. What is the preconstruction groundwater regime as a function of depth:

   a. Is it hydrostatic, that is, with insignificant gradients of pressure head?

   b. Are there artesian pressures?

   c. Are there perched water levels, leakage or infiltration that will cause concentrated shallow flow to the excavation?

2. What changes to the ground water regime will occur during and after excavation and construction?

3. What effects will these changes in groundwater regime have on:

   a. Design and performance of the excavation bracing system?

   b. Design and performance of the foundations to be constructed within the excavation?

   c. Performance of the foundations of adjacent properties?

3.4 Specification Requirements for Water Control

Both "method" and "performance" specifications are used in dewatering practice. Performance specifications are much more common, mainly because control of groundwater is such a specialized activity that much of the expertise resides with contractors.

3.4.1 Performance Specifications. These may stipulate some or all of the following items:

1. Designation of piezometric levels to be achieved by drawdown, generally for locations beneath the interior subgrade.

2. General requirements for drying of interior berms or excavation cut slopes.

3. Qualifications for the dewatering subcontractor by experience and job record.

4. Requiring a subcontractor's proposal for a guaranteed "dry job".

5. Basic technical requirements for the dewatering system, such as the minimum number or maximum allowable spacing of dewatering units or of pumping capacity.

6. Provisions for flooding the excavation in an emergency caused by excessive inflow of seepage.

7. Requirements for continuing the dewatering effort while the permanent structure is being built and concrete is curing.

8. Diversion of surface water and sealing, supporting, strengthening, or re-routing of utilities.

9. Restrictions on the depth of exterior drawdown and requirements for recharge to maintain exterior water levels.

10. Requirements for standby or backup pumps and equipment in case elements of the dewatering system fail.

3.4.2 Method Specifications. These usually develop from the special field tests and investigation of Section 3.3. A minimum dewatering plant is specified in some detail, calling for a basic bid price for the stipulated system with provision for extra payment for installation of additional units. It may be applicable under some of the following circumstances:

1. Exceptional size of the project or potential difficulty with the excavation.

2. Special expertise or experience of the owner's engineering staff.

3. The threat of danger to the project or damage of adjacent property and facilities.

4. Proven economic or technical advantages of a certain dewatering method.

3.5 Maintaining a Dewatered Excavation

The need for maintaining a dewatered excavation will depend on subsoil conditions and the type of construction intended within the retaining system. Under special circumstances, it may be advantageous to temporarily maintain a full head of water within the excavation and place tremie concrete to seal the bottom. The concrete seal can also be designed to act as a structural member resisting hydrostatic uplift from below and lateral loads.

It is good practice to maintain the bottom of the excavation free of standing water. The presence of free water, together with the reduction in total vertical stress that accompanies excavation, can lead to swelling and softening of bearing soils at subgrade. Disturbance that results as personnel and equipment move over the softened subgrade can lead ultimately to detrimental settlement of the permanent foundation. The potential for disturbance generally decreases with increasing permeability and grain size of the subsoil. Perhaps the most critical situation occurs with silty fine sands or silts of low plasticity and little cohesion. Problems are greatly aggravated by the presence of a source of artesian pressures or high seepage flow at a shallow depth below subgrade. A particularly unpleasant condition is created by a concentration of flow in pervious material above an impermeable stratum when this interface is intersected above the base of excavation. It may be that no practical dewatering system can prevent concentrated seeps into the excavation at low points on this impervious surface.

3.6 Controlling Surface Water

Surface water entering the excavation can be troublesome and dangerous. In addition to the threat of flooding, it can erode exposed slopes with possible loss of stability and consequent silting of the bottom of the excavation. The underdrain system of the permanent construction may also be adversely affected. Surface water can originate as runoff from adjacent properties, backup from storm sewers and leakage from adjacent utilities. Converted by infiltration through pervious surface strata or tension cracks to a perched water table, it can apply additional pressures on elements of the bracing system. At a northern location this infiltration can aggravate development of ice lenses in frost-susceptible soils with attendant ice pressures. These pressures can apply a "following" load of high local intensity.

26

Various means of controlling surface runoff are appropriate, including exterior dikes and ditches, or extending the support wall of the excavation above the outside grade. Significant runoff should not be allowed to pond outside the excavation since it may gradually infiltrate the ground and build up water pressures against a tight retaining system. Specifications should inform bidders of known potential sources of difficulty and charge the contractor with responsibility for taking protective measures for runoff control.

Overflow from storm sewers and leakage from adjacent utilities whose condition is either pre-existing or due to construction damage cannot necessarily be predicted. However, appropriate preconstruction assessment of the location and condition of sewers and other utilities could lead to proper precautions, such as necessary re-routing and repairing of utilities or monitoring of performance during construction so that flow can be halted or diverted. The threat to a utility posed by the excavation depends on the stability of the sides of the excavation and the influence zone of potential movements. When an evaluation indicates potential danger to utilities, rerouting or supplementary support should be considered.

3.7   Controlling Groundwater

The ability to control groundwater is related to the following factors: available pumping capacity, differential head, geometry of the excavation, type of excavation support system (particularly the type of wall), subsoil stratification and permeability, and local conditions of recharge. Local recharge from sources such as nearby open water, leaking utilities, surface drainage or irrigation, may be difficult or impractical to control. If potential troubles from recharge have been corrected, subsoil conditions and the general scale of the excavation will control selection of the dewatering system. Drawdown may have to be continued during the early stages of permanent construction. The latter may require temporary underdrains, bleeder pipes through slabs on subgrade and the eventual grouting or sealing of this temporary system. The role of these various factors is discussed below.

3.7.1 Available System Capacity. Theoretically, pumping capacity can be increased to almost any desired level merely by supplying a sufficient number of pumps. However, it is often uneconomical to arbitrarily increase the number of pumps because of constricted space, limitations on the collection systems, and differing efficiencies of various pump types. For high flow quantities the header capacity often can throttle the discharge rate.

3.7.2 Differential Head. The quantity of seepage toward an excavation is related to the differential head. Usually, there is no means of varying the differential head since it is fixed by existing groundwater conditions, depth of the excavation, and hydrology of the area. However, it is essential to estimate the range of differential head that may occur during construction so that the dewatering system may be designed accordingly. In many cases it is necessary to take into consideration the seasonal high groundwater levels.

3.7.3 Excavation Geometry. Figure 3-1 identifies significant features of the excavation geometry in relation to groundwater control. Seepage can enter the excavation through the walls and bottom. If the excavation is wide compared to the depth of the circumferential cutoff wall or if artesian pressures are present in deeper strata, the seepage quantities may be significant. Seepage quantities should be evaluated before construction, using appropriate flow equations or by means of flow nets or other analytical techniques. Computer software for analysis of general seepage problems are available. Often, the flow can be analyzed with simple formulas found in Reference 3-1 by considering the excavation as a large well if the ratio of excavation depth to width is close to unity, or by considering the excavation as a slot if the ratio of excavation depth to width is large. The accuracy of the seepage analysis is limited less by theoretical considerations than by knowledge of subsoil permeability. The margin of error in the theoretical analysis typically will be less than 50 percent. Errors introduced by uncertainties in the permeability could vary over one order of magnitude. Selected system capacities typically should be two to five times the value of a conservatively computed seepage flow.

3.7.4 Excavation Support System. Seepage through the wall should be assessed on the basis of differential head, retained soil characteristics, and effective permeability of the wall. Unfortunately, the effective permeability of the wall is mainly controlled by defects resulting from construction. For example, a torn or jumped interlock of a steel sheet pile wall could allow an order of magnitude more seepage through the wall (and movement of fines) than indicated by laboratory or theoretical studies of seepage through "tight" interlocks. Similarly, a local zone of honeycombed concrete could allow significant seepage compared to properly placed concrete. Voids created in the soil outside the wall by installation procedures such as jetting or pre-augering can greatly aggravate leakage through the wall. Some wall systems such as soldier piles with lagging are intended to be so pervious as to eliminate water pressures on the back of the wall. Qualitative comparisons for various types of well-constructed walls are shown in Table 3-1.

3.7.5 Subsoil Conditions. The single most important factor in establishing a suitable means of removing groundwater from an excavation is the subsoil permeability and stratification. The steady state seepage quantity is directly proportional to the soil's coefficient of permeability. Estimates of permeability based on laboratory tests are frequently in error by an order of magnitude or more. For clean granular-materials with no more than about 5 to 10 percent passing the No. 200 sieve size, correlations of permeability with grain size parameters are often sufficiently accurate (Reference 3-1), Where marked anisotropy and pronounced strata changes are present, field pumping tests may be necessary to

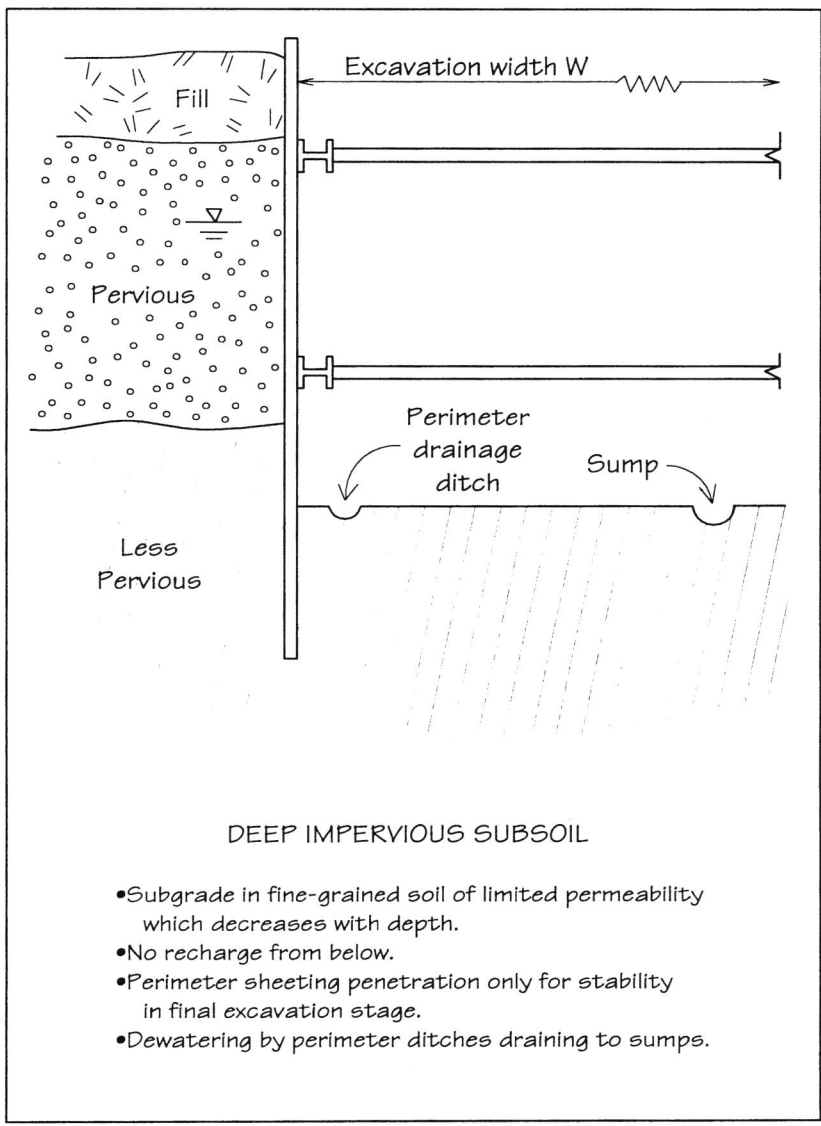

DEEP IMPERVIOUS SUBSOIL

- Subgrade in fine-grained soil of limited permeability which decreases with depth.
- No recharge from below.
- Perimeter sheeting penetration only for stability in final excavation stage.
- Dewatering by perimeter ditches draining to sumps.

Figure 3-1a
Significant Features of Excavation Geometry
in Relation to Groundwater Control

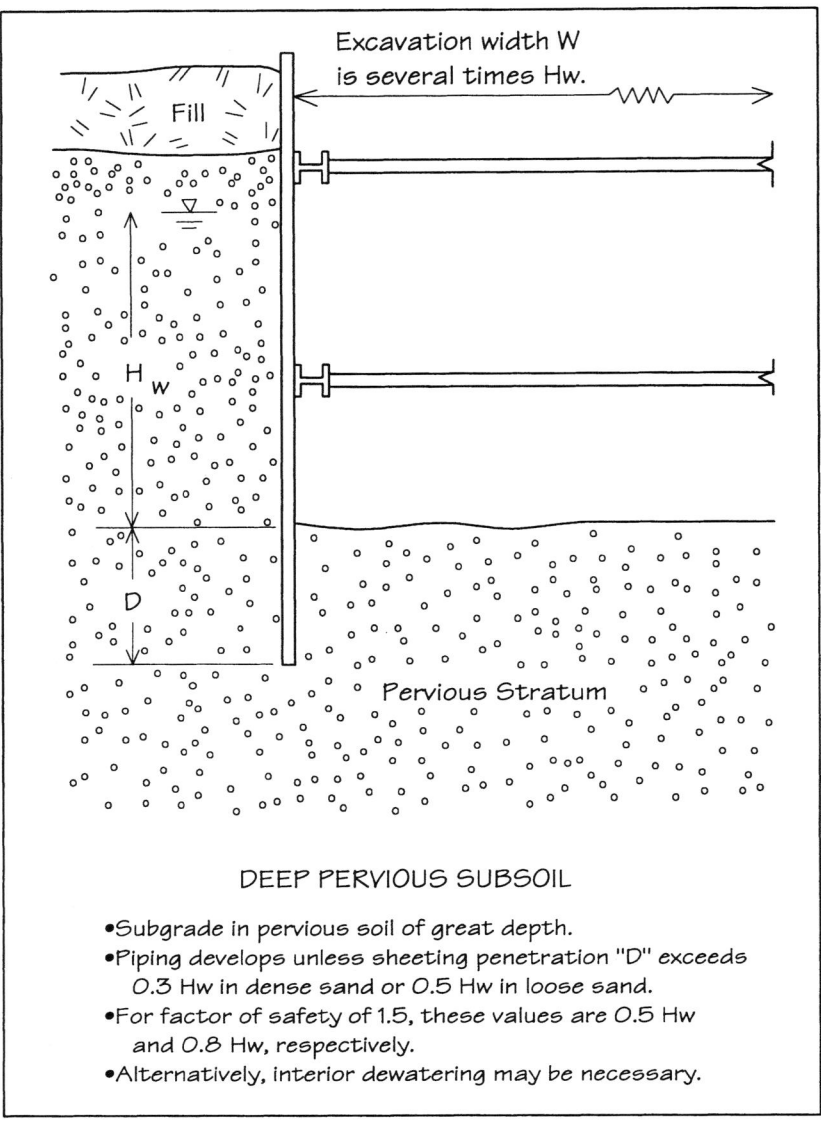

Excavation width W is several times Hw.

DEEP PERVIOUS SUBSOIL

- Subgrade in pervious soil of great depth.
- Piping develops unless sheeting penetration "D" exceeds 0.3 Hw in dense sand or 0.5 Hw in loose sand.
- For factor of safety of 1.5, these values are 0.5 Hw and 0.8 Hw, respectively.
- Alternatively, interior dewatering may be necessary.

Figure 3-1b
Significant Features of Excavation Geometry
in Relation to Groundwater Control

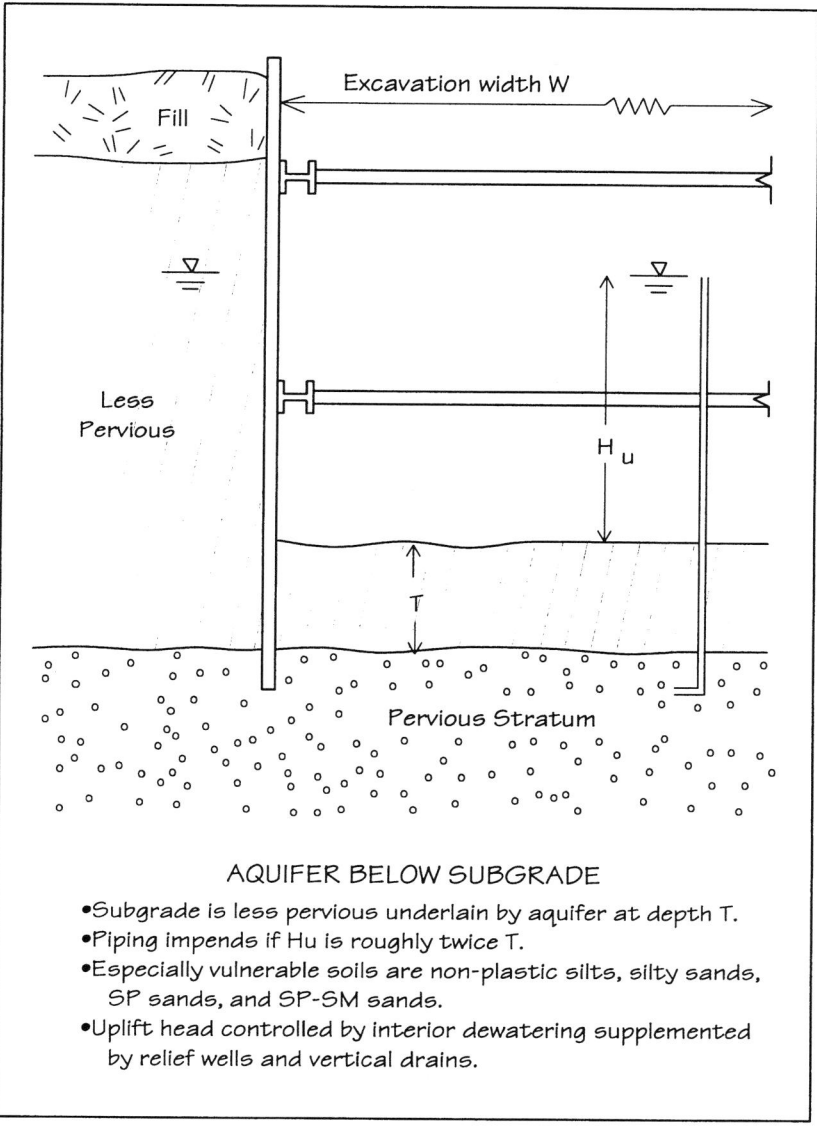

AQUIFER BELOW SUBGRADE

• Subgrade is less pervious underlain by aquifer at depth T.
• Piping impends if H_u is roughly twice T.
• Especially vulnerable soils are non-plastic silts, silty sands, SP sands, and SP-SM sands.
• Uplift head controlled by interior dewatering supplemented by relief wells and vertical drains.

Figure 3-1c
Significant Features of Excavation Geometry
in Relation to Groundwater Control

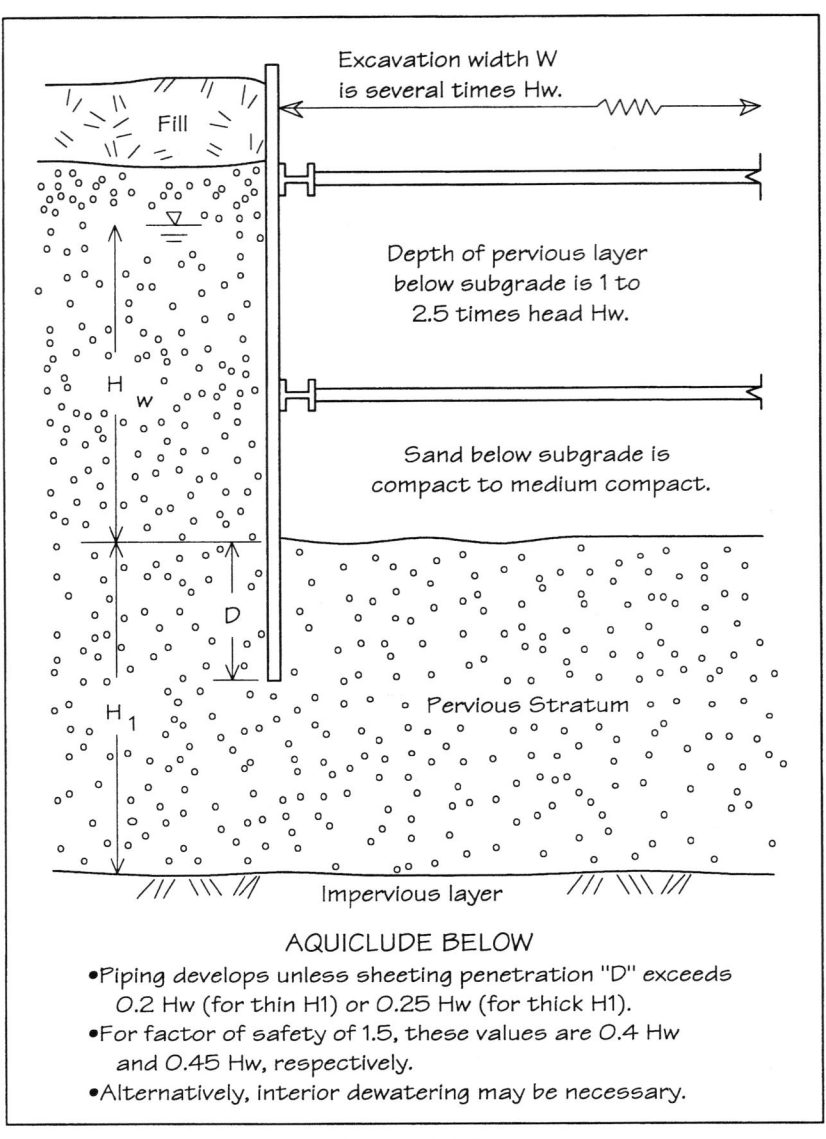

Excavation width W is several times Hw.

Depth of pervious layer below subgrade is 1 to 2.5 times head Hw.

Sand below subgrade is compact to medium compact.

Fill

Hw

D

H1

Pervious Stratum

Impervious layer

AQUICLUDE BELOW

- Piping develops unless sheeting penetration "D" exceeds 0.2 Hw (for thin H1) or 0.25 Hw (for thick H1).
- For factor of safety of 1.5, these values are 0.4 Hw and 0.45 Hw, respectively.
- Alternatively, interior dewatering may be necessary.

Figure 3-1d
Significant Features of Excavation Geometry
in Relation to Groundwater Control

TABLE 3-1. Cutoff Functions and Watertightness of Excavation Support Walls

TYPE OF WALL	POTENTIAL LEAKAGE THROUGH WALL	POTENTIAL FOR PIPING BENEATH WALL
Wood, concrete, or steel cantilever sheeting of low height	Deflection opens joints between sheets. Leakage typically 0.1 to 4 gpm per 100 linear feet for low head. Also depending on joint detail.	To avoid piping, penetration below subgrade must exceed 1/4 of exterior head where there is limited depth of pervious layer or more than 1/2 of head for deep pervious layer.
Braced, interlocked steel sheeting	Leakage typically 1 to 10 gpm per 100 lin. ft. of head. Lesser quantity of leakage if movement is minimized, locks are filled and tensile stress acts along the wall in longitudinal direction	As above, but in potentially "running" soil - non-plastic silt, silty fine sand or narrowly graded sand - piping may occur in a path along the face of the sheeting.
Steel soldier piles with wood or concrete lagging	No impediment to leakage, exterior drawdown and water pressure in active wedge of retained soil can be analyzed by flow net. Inflow depends on soil permeability and recharge.	Not intended for cutoff below subgrade. If exterior head is not greatly reduced by leakage, dewatering to control uplift may be necessary.
Concrete cylinder piles: tangent, secant, or staggered	Equivalent permeability of wall is typically 5×10^{-4} to 10^{-5} cm/sec. Leakage typically 1/2 to 8 gpm per 100' per 10' head, depending principally on quality of joints between cylinders.	Could accomplish partial of complete cutoff below subgrade, but usually not seated into rock. Layout convenient to work around shallow obstructions.
Slurry trench concrete wall	Equivalent permeability of wall is typically 2×10^{-4} to 10^{-6} cm/sec. Leakage typically 0.1 to 4 gpm per 100' per 10' head. Much influenced by tieback penetration through wall and joint quality. Details of penetrations are important.	Cutoff enhanced by chiseling into underlying rock of by prepositioning grout pipe in wall elements to facilitate grouting in strata beneath wall.

Note: (1) Leakage through wall is expressed as quantity of inflow per 100-foot length of wall for each 10-foot difference in head across the wall, assuming leakage is not limited by permeability of the retained soil. (2) 1 foot = 0.3 meters; 1 gpm = 6×10^{-5} m^3/sec.

evaluate in-situ permeability. Sometimes, falling or rising head borehole permeability tests at various elevations will provide sufficient information for design. However, these tend to yield lower values of permeability than pumping tests because it is difficult to keep the most pervious materials open for testing in the walls of the boreholes.

3.7.6 <u>Methods of Achieving Drawdown</u>. Methods of lowering groundwater level by removing the water include:

1. Deep wells

2. Single-stage wellpoints

3. Staged wellpoints

4. Eductor wellpoints

5. Deep sheeted sumps and pumps

6. Shallow ditches, sumps and pumps

7. Relief wells

The range of applicability of these methods to groundwater removal is summarized in Table 3-2. In many construction situations, a combination of two or more elements may comprise the complete system. Electro-osmosis is a special method of reducing piezometric levels and strengthening silts and clays.

Alternatively, methods of creating an impervious barrier by chemical or cement grouting, soil freezing, or selected retaining wall systems can be utilized to minimize or eliminate the dewatering effort. Chemical grouting in particular is now often used to reduce the total quantity of inflow or to control concentrated seeps.

3.8   <u>Reducing Water Pressures on the Support System</u>

Reducing water pressures that act on the support system can improve the economy and safety of the system. However, other problems, as discussed below, may be introduced by measures which reduce external water pressures. The simplest method of draining the retained soil is by use of an open type, or "leaking" wall, such as a soldier pile-lagging wall. This method is applicable when the resulting inflow can be conveniently handled by pumping from shallow sumps, and when migration of fines through the wall is effectively prevented or has no serious consequence. If piping of fines is a problem, it is usually minimized by packing

34

TABLE 3-2. Methods of Ground Water Control

CONTROL METHOD	SOIL TYPES IN WHICH MOST SUITABLE	CONTROL METHOD MAJOR ADVANTAGES	MAJOR DISADVANTAGES
(1) Deep Wells	Deep homogeneous deposits with permeability greater than about 10^{-4} to 10^{-3} cm/sec. Also highly stratified and layered deposits or where permeability increases with depth.	Large areas can be dewatered with individual wells; discharge system can be routed to minimize interference with construction; deep drawdown and large lifts can be obtained. Facilitates pre-drainage before excavating.	Relatively expensive installation costs; required adequate thickness of aquifer to develop drawdown; may require significant time to drawdown the water table; good filter, casing and pump design required.
(2) Single Stage Wellpoints	Deposits with permeability greater than about 10^{-4} cm/sec. More variability, more highly stratified, or more recharge than (1).	Easily and rapidly installed; spacing can be readily altered to suit local conditions; sanding of entire wellpoint hole can reduce effects of stratification and is common and conventional.	Limited to approx. 15 ft. water lift; requires fairly close spacing of well points (usually less than 10 ft.); discharge system may interfere with construction; relatively high operating costs.
(3) Staged Wellpoints	Same as (2) but can accommodate artesian conditions and upward gradients.	Same as (2); can handle larger water lifts, but in 15 ft. stages.	Same as (2); construction interference may become very significant.
(4) Eductor Wellpoints	Same as (2) but generally appropriate for less permeable material or more distinctly stratified material.	Same as (2) but higher lifts can be achieved. Useful for predrainage.	Same as (2) (except for lift limitation); significantly more expensive; requires more tuning, maintenance & technical attention.
(5) Deep Sheeted Sumps with Pumps	Same as (1) but where conditions are simpler and fewer units suffice.	Same as (1) but less flexibility and adaptability to uncertain ground conditions.	Same as (1) but more expensive; careful design of filters required to prevent loss of ground by piping into sump.
(6) Shallow Ditches, Sumps and Pumps	Permeability and stratification less important. Inappropriate where artesian conditions or large upward-directed gradients are present.	Easy and inexpensive to install, no special skills, no specialist subcontractor required.	Limited drawdown produced; discharge system may interfere with construction; filter design is important; may require significant time for dewatering and can result in difficult excavation and bottom conditions.
(7) Relief Wells	To penetrate pervious stratum capped by less pervious layer. Usefulness generally limited to special situations.	Can minimize potential for blow-out, used in combination with pumping wells. Relatively simple and inexpensive to install and maintain.	Not adaptable for pre-drainage; drawdown is limited by elevation of discharge from well. Usually required to be used in conjunction with other procedures.

35

the spaces between lagging with straw, salt hay, gravel, oakum, or similar filtering material.

Other methods utilize a wellpoint or deep well system around the excavation or by pumping from considerable depths below the base of excavation (with wells, sumps or wellpoints) so that a large cone of depression is developed. The latter method usually requires more time to become effective.

The most serious problem resulting from reducing water pressures on the support system is the more widespread lowering of water levels at distances from the excavation. This is important if structures situated within the drawdown zone are subjected to settlements. Methods of evaluating this problem are discussed in Section 3.11.

3.9   Minimizing Disturbance of the Excavation Bottom

The main cause of disturbance of fine-grained soils is the movement of personnel and equipment over a subgrade that has softened due to removal of overlying weight and the action of seepage. In coarser soils an upward seepage gradient causes a reduction in effective stress, and thus, a reduction in shear strength and an increase in compressibility. Disturbance is minimized by placing a crushed stone or gravel ballast layer on the subgrade and pumping water from sumps or ditches in the ballast. For fine-grained soils, a "mud mat" of lean concrete acts as a pavement, distributing the concentrated weight of personnel and equipment. The mud mat may have to be drained by bleeder holes connected to gravel pockets beneath the mat. Alternatively, a filter fabric can be placed on a silt or clay subgrade beneath stone ballast. For coarser-grained soils, the ballast acts as a pavement and as a weighted filter, increasing the effective stresses acting on the subsoils. An alternative procedure, which alone may suffice to stabilize coarse-grained soils, is to simply maintain water levels several feet below the base of excavation by pumping.

3.10   Preventing Bottom Blowout

The problems of bottom blowout, boiling, or heave can be aggravated when relatively impervious soils at the excavation subgrade are underlain by pervious strata. Heave can destroy the bearing qualities of the subgrade and increase settlements of the permanent structure. Therefore, this possibility must be addressed by the owner's engineer in framing specifications and by the contractor in his project planning.

In relatively homogeneous pervious soils, a similar problem can develop from upward seepage beneath a tight cutoff wall causing boiling of the bottom or a "quick condition". Conditions causing piping through the subgrade are illustrated in Figure 3-1. Heave can be controlled by wells, wellpoints, or deep sumps which reduce water pressures below the base of excavation. Difficulties often can be lessened by increasing the penetration of the wall. However, it may be undesirable to drive sheet piling through a silt or clay stratum at subgrade to a partial penetration into underlying sands since a ready path for piping can be created along the face or the sheet piling. Reference 3-4 provides simplified procedures for evaluating the required penetration depth.

If bottom blowout or boiling occurs, it can lead to complete failure of the bracing system by causing loss of toe support for the walls. Therefore, it is of utmost importance to identify this problem before construction. This requires detailed identification of soil strata and piezometric levels at and just below the base of excavation. Its control will require monitoring of piezometric levels below the excavation during construction, as discussed in Chapter 8.

3.11 Effects of Dewatering on Adjacent Areas and Control Measures

Dewatering can result in distress to adjacent properties manifested in settlement caused by the following factors:

1. Piping of fines resulting in loss of ground.

2. Consolidation of compressible soils due to increase in effective stresses.

3. Deterioration of existing foundation elements.

Movement of fines can be minimized by use of properly designed filters at critical water discharge points into the collection system. "Leaking" walls are most susceptible to this type of distress. Movement of fines can also occur upward through the base of excavation. Properly selected filters which will pass seepage but restrain the fine grain sizes of the potentially erodible soils can minimize the problem. If the rate of excavation is slow, the water pressures outside the wall or beneath the subgrade may be gradually reduced to the extent that a major dewatering effort is unnecessary.

Lowering of groundwater levels outside the excavation can cause significant settlement if compressible strata are present. The settlement magnitude is a function of soil compressibility and of the pore pressure reduction which causes an equal increase in effective stress. Compressibility is chiefly influenced by the degree of preconsolidation of the subsoil. A reliable profile of subsoil preconsolidation stresses is essential to a realistic settlement prediction. Generally, dewatering causes insignificant compression of heavily overconsolidated soils, but can be significant in normally consolidated or slightly overconsolidated soils. The profile of settlements produced in a direction away from the excavation usually is smooth and gradually sloping, reflecting the flat slope of the drawdown curve. However, noticeable differential settlement can occur at the contact between pile-supported and ground-supported elements.

Old foundations, especially untreated timber piles, can deteriorate rapidly if groundwater lowering exposes them to oxygen enrichment. Negative skin friction, developing as a result of consolidation of soft soils around piles driven through them to firm bearing, must also be considered. Ancillary problems such as the interception of a neighbor's well water supply, and either the real or imagined depletion of water utilized by plantings should be identified and taken into account in framing specification requirements which limit drawdown or dewatering efforts. Concern for trees and shrubs may necessitate a specification requirement for watering them at the surface.

The following measures can be utilized to control detrimental effects of exterior drawdown:

1. Creating a positive cut-off with a deep wall to maintain pre-existing exterior water levels.

2. Recharge systems to pump the water back into the soil or to raise the piezometric levels. Such systems have met with varying degrees of success. A recharge system will be ineffective unless an overpressure is used in the recharge well to raise piezometric levels. The maximum overpressure in the system is limited by the overburden weight and soil strength which set the value at which hydraulic fracturing will occur. Also, the high water pressures in the recharge system could cause unexpected loadings and possible distress of the excavation support system.

3. Careful monitoring of the construction, including vertical and horizontal movements of the adjacent ground surface and any structures. For the monitoring to be successful, however, mitigation plans and equipment should be available for responding to problems.

3.12 Monitoring Groundwater Changes

Construction monitoring of groundwater changes includes the following basic activities:

1. Measurement of the flow quantities either discharged from or recharged into the soil.

2. Sampling and testing of the effluent to determine the presence of fine particles of soil, sewage, or chemicals.

3. Observation of groundwater levels (or pore water pressures in relatively impervious soils) by means of installed piezometers.

3.12.1 Quantity of Flow. The quantity of flow should be monitored to verify that pumping capacity and dewatering system design are adequate. Often a simple weir or measurement of the trajectory of the stream as it flows from a discharge pipe is all that is necessary. For other situations, accurate flow meters may be required. Sometimes several monitoring stations may be necessary to isolate and evaluate critical sources of seepage.

3.12.2 Quality of the Effluent. At the least, periodic visual observations should be made to ascertain the presence of fines (indicating possible subsurface erosion) or sewage (indicating damage to sewer lines). Depending on the circumstances, quantitative measurements of these factors may be necessary. Chemical quality determinations may be advisable, especially where strata contain soluble calcareous or gypsum constituents.

3.12.3 Groundwater Levels. Groundwater levels (or pore water pressures) should be monitored where it is necessary to verify design assumptions, particularly with regard to hydrostatic forces acting on the excavation support system or potential uplift of the subgrade. Sometimes, nothing more than visual observation of the breakout level of seepage is required. For other situations, detailed measurements with observation wells and piezometers are necessary. Further discussion of the role of water level observations is included in Chapter 8.

3.13   References

3-1   Mansur, C. I., and Kaufman, R. I. (1962). "Dewatering," *Foundation Engineering*, Ed. by G. A. Leonards, McGraw-Hill Publishing Company, New York, NY, 241-350.

3-2   Powers, J. P. (1981). *Construction Dewatering: A Guide to Theory and Practice*. John Wiley & Sons, New York, NY.

3-3   Groundwater Committee of the Underground Technology Research Council (1985). *Dewatering-Avoiding Its Unwanted Side Effects*. Ed. by J. P. Powers, ASCE, New York, NY.

3-4   Naval Facilities Engineering Command (1986). "Soil Mechanics: Design Manual 7.1," *NAVFAC DM-7.1*, Department of the Navy, Alexandria, VA, 259-307.

CHAPTER 4: SYSTEM STABILITY AND WALL LOADING

4.1 Fundamental Considerations

Earth loads and support system stability are basic considerations in design of earth retaining structures. In this chapter, methods of analysis of loadings and stability of braced and tied-back walls will be examined. Empirical modifications to the methods will be explained. Further recommendations and their use in design are given in Chapters 5, 6, and 7.

4.1.1 System Stability. In this discussion, excavation stability is a term applied to problems of potential shear failure in the soil mass retained by the wall system or in front of the wall system. Stability of the excavation is a prime consideration in the initial decision to provide support. If the excavation is potentially unstable, then a wall must be designed to help restrain it. In such cases, the support loads are, in large part, a function of the degree of instability of the excavation.

A conventionally designed structural system does not necessarily eliminate all stability problems. Several modes of soil failure can occur in spite of the support, as illustrated in Figure 4-1. In Figure 4-1 (a), a slip in a berm above the wall could fall into the excavation. Failure in a temporary internal berm as in Figure 4-1 (b) can lead to collapse of the wall. Figure 4-1 (c) depicts a potential failure mechanism for soil over rock; daylighted joints in the rock lead to slippage of the rock from beneath the wall system itself. Deep-seated failures as shown in Figure 4-1 (d) and (e) lead to heave in the excavation bottom. Failure behind the anchors of a tied-back wall, as in Figure 4-1 (f), results in collapse of the wall system. All these failure modes need to be investigated in design.

4.1.2 Concepts of Earth Loading. Loads produced on a support system are a function of system geometry, soil properties, stratification, seepage conditions, construction sequence, system stability, environmental factors, surcharge loads, prestressing loads, and movements. Each of these items must be taken into account in selecting design loadings.

In considering earth pressure loadings, two basically different approaches may be followed:

1. The bracing system is designed to carry only the load exceeding that which the earth mass itself resists. In this case, the movements needed to mobilize soil strength are considered acceptable. Design loadings are based on active or near-active case wherein the soil mobilizes its full strength along some potential failure surface. Pressures on the wall are not necessarily distributed in a triangular diagram over its height.

40

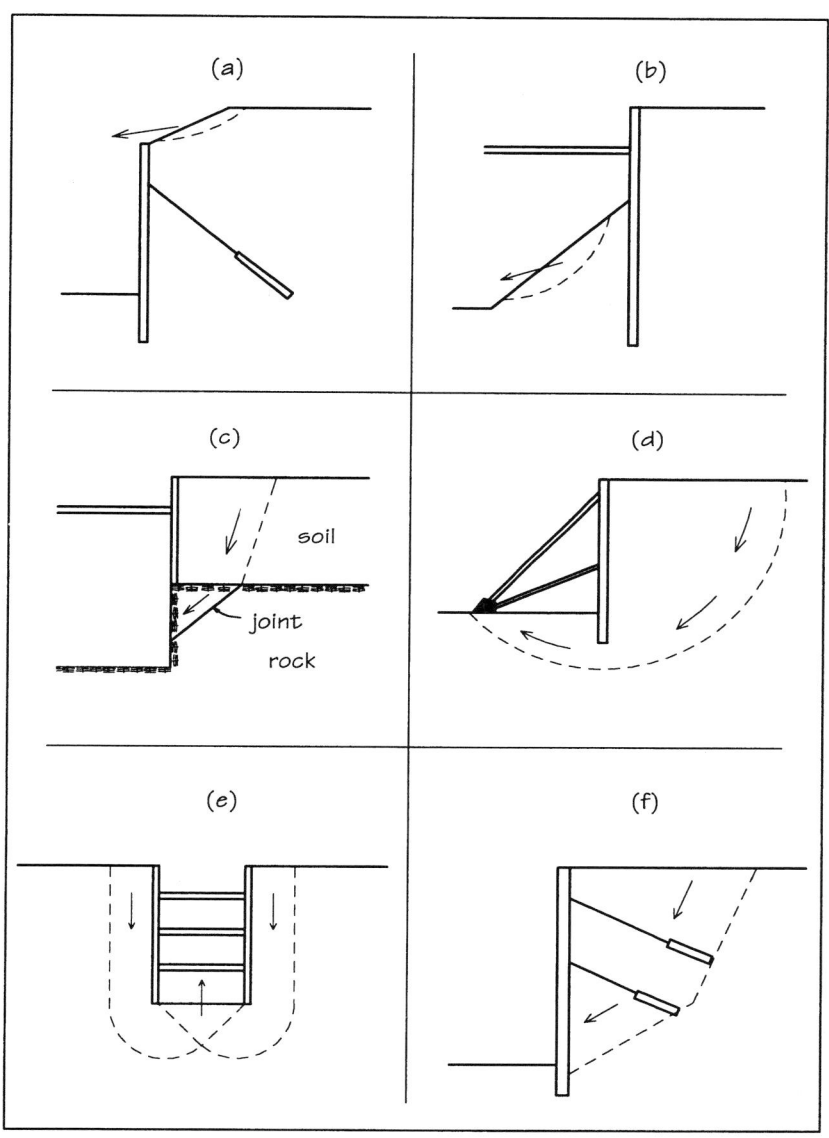

Figure 4-1
Modes of Stability Failures

41

2.   Alternatively, if minimizing deformation is mandatory, then the bracing system may be designed to be stiff and/or highly preloaded. This requires that the structural system carry much of the load which would otherwise be resisted by the soil itself. Loadings on the structure are then greater than for the active case.

4.1.3 Relationship Between Movement and Earth Loads. The sequence of deformation of a conventional braced wall as excavation proceeds is illustrated in Figure 4-2 (a). The pattern of movement of the wall is dominated by rotation of the bottom of wall about the uppermost strut. If deformations are large enough, this leads to mobilization of soil strength along a potential failure surface as shown in Figure 4-2 (b). Equilibrium of the mass of soil above the failure surface is established by an external force, P, which is the resultant load acting on the support system.

4.1.4 Preloaded Versus Unpreloaded Systems. Preloading a wall system may change the location and magnitude of the net resultant; the effect depends upon the magnitude and distribution of preload forces. If a wall is preloaded with a total force less than the active resultant, it will behave essentially as an unpreloaded wall because in both cases, the restraining forces are not adequate to prevent wall movement and the consequent full mobilization of soil strength along a potential failure surface.

To prevent full mobilization of soil strength with accompanying strains, preloading forces must total more than the active resultant and the moment of preloading forces about the failure surface rotation center must exceed that of the active resultant. These forces will then significantly influence the location and magnitude of the earth loading resultant. Other factors such as degree of mobilized soil strength, wall flexibility and prestress load spacing also play a role. Inward wall movements, due to excavation stress relief, are, in part, counteracted by outward wall movements during preloading.

4.2   Stability Analysis

Stability considerations often play an important role in design of the support system. In this section, techniques for analysis of potential slope and base failure are summarized. Potential application of the techniques in determining earth loadings are deferred to Section 4.4, but it should be noted that wall loading can be controlled by stability considerations.

4.2.1 General Trial Solutions. Trial stability solutions are the most general of stability analysis tools. In a trial technique, a plausible failure surface is assumed and a limiting equilibrium analysis of the soil mass above this surface is conducted so as to calculate the safety factor provided by a certain design. After a number

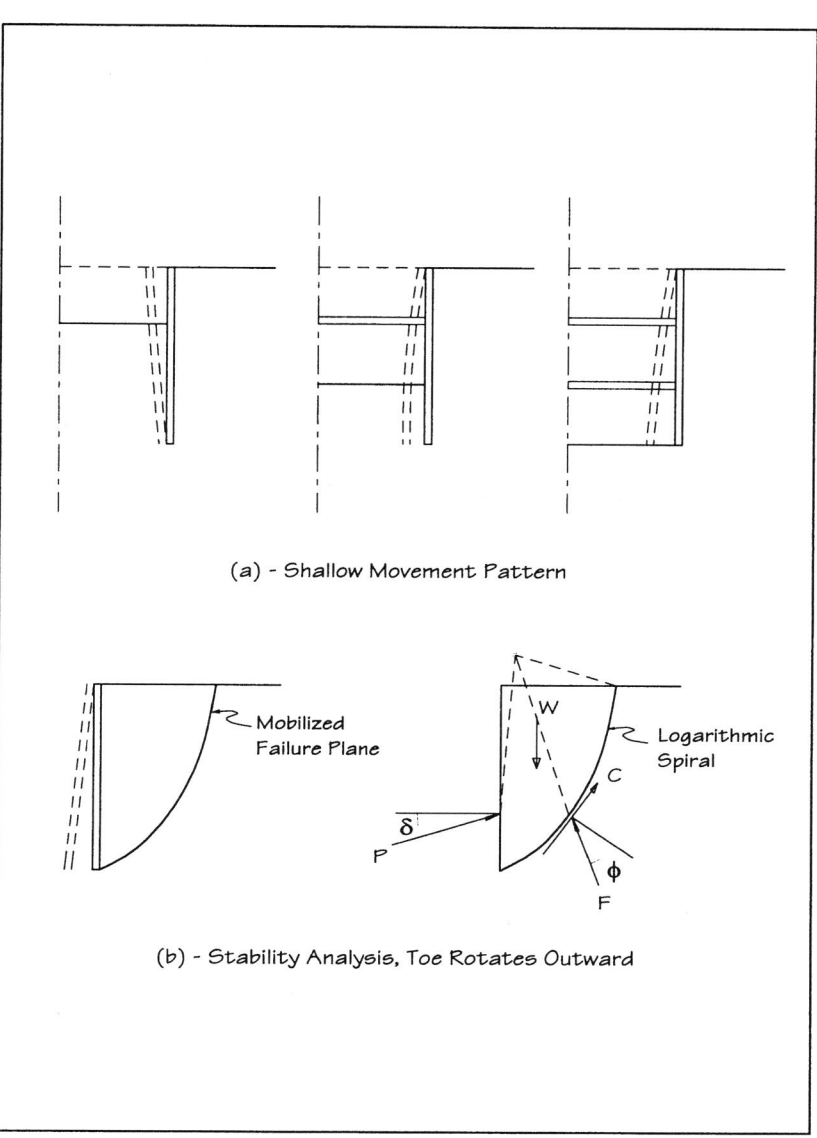

(a) - Shallow Movement Pattern

Mobilized
Failure Plane

W

Logarithmic
Spiral

C

δ

P

φ

F

(b) - Stability Analysis, Toe Rotates Outward

Figure 4-2
Modes of Braced Wall Deflections

of such trials, a minimum safety factor is obtained for the critical surface. These methods are useful in analysis of almost any of the possible modes of failure illustrated in Figure 4-1.

Table 4-1 lists common methods of stability analysis with comments on the capabilities and advantages of the methods. Of these techniques, much more experience has been acquired with the "$\phi = 0$" analysis, the ordinary method of slices, the Modified Bishop procedure, and the wedge method than with the remaining . These established procedures are suitable to handle most stability analyses for temporary wall systems. In some specialized cases, the newer and more general Janbu or Morgenstern and Price methods might be preferable. A number of available computer programs facilitate the use of both older and new procedures. The Jelinek and Ostermayer procedure, illustrated in Reference 4-11, is designed specially to analyze the failure possibility of tieback walls.

4.2.2 Stability Charts. For certain idealized conditions, stability charts have been developed using the trial procedures to facilitate stability evaluation. A summary of published charts is given in Table 4-2 with comments as to types of problems covered by each chart. Before a trial solution is undertaken, the charts should be consulted to determine if a solution is not already available or if useful guidance could not be obtained for the trial analysis.

4.2.3 Vertical Unsupported Slopes in Cohesive Soils. Simple solutions for the critical height at the limit of stability of a free-standing vertical face are contained in many textbooks. For the case of cohesive soils with $\phi = 0$, the expression for critical height, H_c, is $(4c)/\gamma$, where c is the undrained shear strength and γ is the soil total unit weight. Unfortunately, a slope which stands at this critical height is in a state of precarious equilibrium. This expression for critical height assumes that tension can exist in the upper portion of the slope, no cracks occur in the soil, no flow of water will take place, no surcharge is placed adjacent to the cut and the soil can be characterized as having constant cohesive strength during the life of the slope.

Because such assumptions are rarely realistic and because failures in vertical slopes can easily lead to loss of life, it is prudent to provide at least light bracing for theoretically "stable" vertical cuts exceeding about five feet in height. At least one row of struts should be placed near the top of the cut to minimize the effects of tension cracking. Local building regulations and OSHA codes should be consulted in this case since many codes specify the nature of required bracing. According to CSHA, any trench with a depth greater than five feet must be shored, sheeted, braced, sloped or otherwise supported. Design of the support system must meet the criteria of a qualified person, defined as a professional engineer or one who by extensive experience is capable of such design.

TABLE 4-1. Common Methods of Analysis of Support System Stability

METHOD	AUTHOR (S)	ASSUMPTIONS		COMMENTS
		FAILURE SURFACE	MATERIAL BEHAVIOR	
"$\phi = 0$"		Circular	$\phi = 0$	Used extensively for short term stability problems in clays. Verified by many successful experiences.
Ordinary Method of Slices	Fellenius (1939)	Circular	ϕ, c	Used extensively in practice, flexible, applicable to many conditions. Conservative, doesn't satisfy all equilibrium equations.
Modified Bishop Procedure	Bishop (1955)	Circular	ϕ, c	Similar to ordinary method of slices but less conservative. Very useful in practice.
Sliding Wedges		Planar Combinations	ϕ, c	Applicable for analyses of sliding along planes of weak zones. Extensively used in practice.
Generalized Procedure of Slices	Janbu (1957)	Arbitrary	ϕ, c	Useful for analyses of non-circular failure surfaces. Can be solved by hand.
Morgenstern and Price	Morgenstern and Price (1967)	Arbitrary	ϕ, c	Useful for analyses of non-circular failure surfaces. Computer solution required. All equilibrium equations equilibrium satisfied.
Jelinek and Ostermayer	Jelinek and Ostermayer (1967)	Planar Combinations	ϕ, c	Specially adapted procedure for analysis of anchor pullout of tied-back walls.

45

TABLE 4-2. Useful Charts for Stability Analysis

REFERENCES	CONDITIONS ADDRESSED
Janbu (1954) NAVFACS (1986)	"$\phi = 0$" and "ϕ - c". Circular failure surfaces assumed. Total stress analysis. Effects of surcharge, submergence, and tension cracking considered. Very useful in practice.
Kerney (1963)	"$\phi = 0$" where undrained shear strength increases with depth. Circular failure surfaces assumed. Total stress analysis.
Bishop and Morgenstern (1960)	Effective stresses. Modified Bishop solution with circular failure surfaces. Arbitrary pore pressures and various seepage/groundwater conditions can be addressed.

4.2.4 Heave into the Bottom of a Supported Excavation - Required Wall Penetration. Soil on either side of a braced vertical cut acts as a surcharge loading which tends to fail the excavation base as shown in Figure 4-3. If bottom heave takes place, the mass of soil behind the wall moves directly downward as a bearing failure occurs beneath this mass, resulting in soil heave in the failure zone into the cut. This type of failure is limited to silts and clays. Bottom heave in sands should only occur if upward seepage gradients directed toward the excavation become large enough to liquefy the sand. Chapter 3 gives references with design information to avoid piping conditions.

Expressions for safety factor against bottom heave of a fully braced cut in clay are given in Figures 4-3 and 4-4. Allowable height of excavation before failure is a function of the soil's cohesion and unit weight, width of excavation and depth to a rigid base. Where bottom heave is a problem, either the height of the excavation must be limited, the embedded wall section designed to resist the failure, or special construction procedures used. In this situation, factors such as loss of strength due to creep and effects of anisotropy can be significant. Reference 4-4 gives procedures for considering anisotropy in basal heave analyses and guidance on when it should be considered.

4.2.5 Stability of Internal Berm. Earth berms are often left against the retaining wall to help support the wall during excavation. In all too many instances, berms have failed to serve their purpose. Problems can develop with earth berms because of insufficient passive resistance, a slope failure in the berm or deep-seated soil movements in soft clay beneath the berm. In cases where the berm serves as a critical restraint in soft clay soils several special factors should be emphasized:

1.   Equipment can rework the surface of the berm and cause soil to a depth of several feel to lose strength. A shallow surficial slide can ensue and lead to a larger failure or detrimental movements.

2.   A "stable" excavation with a relatively low safety factor against basal heave may be subject to large, deep-seated movements beneath the berm, minimizing its effectiveness. (References 4-3, 4-13)

Finally, the designer should be aware of the importance of sequence of construction when using berm supports. The berm is usually excavated during later stages of work and replaced with struts or rakers. The contractor must be aware of how the berm is to be excavated relative to the sequence of support installation, otherwise too much of the berm volume can be removed before its supporting function is picked up by the structural members.

47

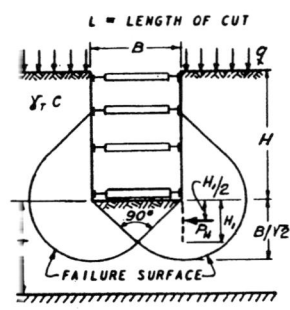

L = LENGTH OF CUT

IF SHEETING TERMINATES AT BASE OF CUT:

SAFETY FACTOR, $F_s = \dfrac{N_c\,C}{\gamma_T H + q}$

N_c = BEARING CAPACITY FACTOR, FIG.12-1, WHICH DEPENDS ON DIMENSIONS OF THE EXCAVATION: B, L AND H (USE H = D).

C = UNDRAINED SHEAR STRENGTH OF CLAY IN FAILURE ZONE BENEATH AND SURROUNDING BASE OF CUT.

q = SURFACE SURCHARGE.

IF SAFETY FACTOR IS LESS THAN 1.5, SHEETING MUST BE CARRIED BELOW BASE OF CUT TO INSURE STABILITY, FORCE ON BURIED LENGTH:

IF $H_i > \dfrac{2}{3}\dfrac{B}{\sqrt{2}}$, $P_N = .7\left(\gamma_T H B - 1.4 C H - \pi C B\right)$

IF $H_i < \dfrac{2}{3}\dfrac{B}{\sqrt{2}}$, $P_N = 1.5 H_i\left(\gamma_T H - \dfrac{1.4 C H}{B} - \pi C\right)$

CUT IN CLAY, DEPTH OF CLAY UNLIMITED (T > 0.7B)

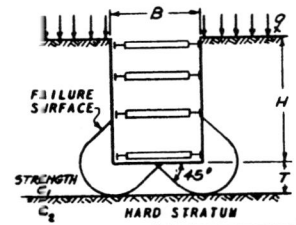

IF SHEETING TERMINATES AT BASE OF CUT

SAFETY FACTOR:

CONTINUOUS EXCAVATION; $F_s = N_{CO}\dfrac{C_1}{\gamma_T H + q}$

RECTANGULAR EXCAVATION; $F_s = N_{CR}\dfrac{C_1}{\gamma_T H + q}$

N_{CO} AND N_{CR} = BEARING CAPACITY FACTORS, FIG. 11-5, WHICH DEPEND ON DIMENSIONS OF THE EXCAVATION: B, L AND H, (USE H = D)

CUT IN CLAY, DEPTH OF CLAY LIMITED BY HARD STRATUM (T ≤ 0.7B)

NOTE: IN EACH CASE FRICTION AND ADHESION ON BACK OF SHEETING IS DISREGARDED. CLAY IS ASSUMED TO HAVE A UNIFORM SHEAR STRENGTH = C THROUGHOUT FAILURE ZONE.

Figure 4-3
Analysis of Basal Heave Potential in Clay Deposits
(reference 4-12)

48

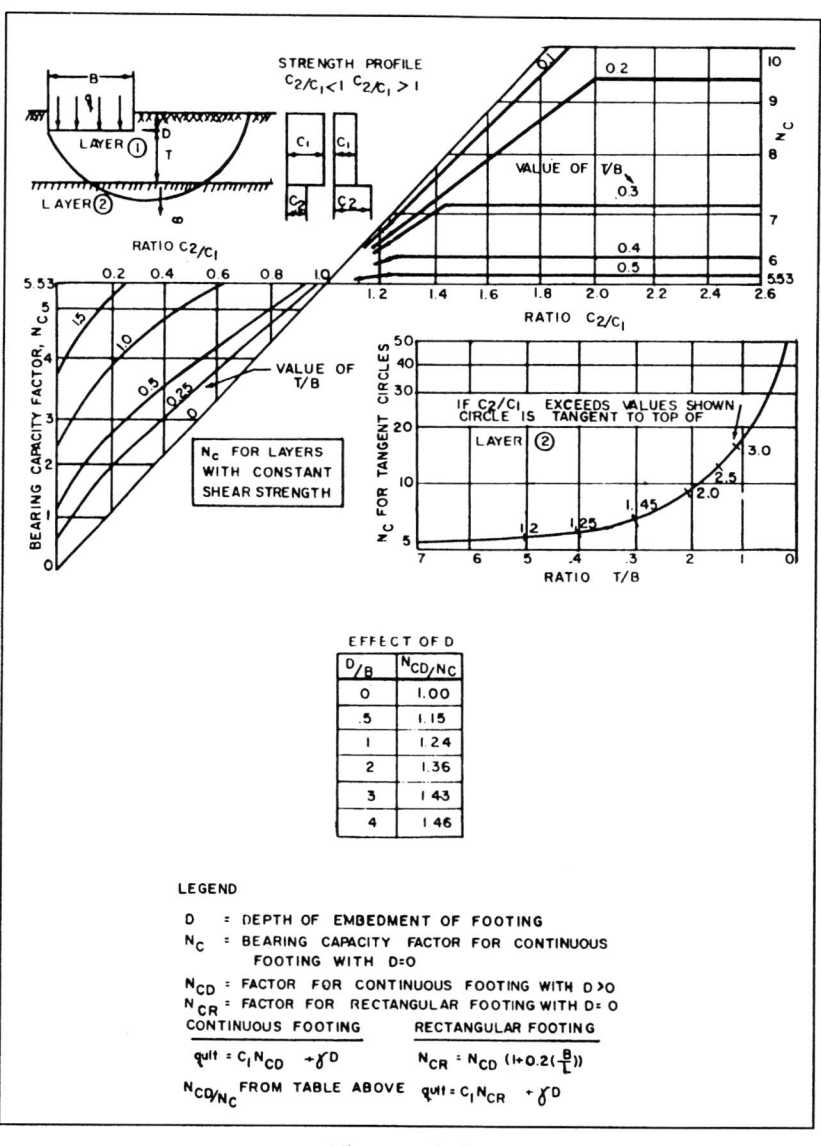

Figure 4-4
Bearing Capacity Factors
for Analysis of Basal Heave Problems
(reference 4-12)

4.3 Design Checklist for Supported Excavation Stability

The preceding section has described methods used to evaluate stability of supported excavations. Various types of potential stability problems were illustrated in Figure 4-1. In this section, a checklist is provided for the designer to review systematically the potential causes of supported excavation instability.

Stability Checklist

1. **Basic Problems**

   (a)   Stability of unsupported vertical or near vertical slopes, Section 4.2.3.

   (b)   Stability of slopes above the excavation which could slide into the cut - use stability charts or trial stability procedures, Sections 4.2.1, 4.2.2.

   (c)   Stability of any internal berms - use stability charts or trial procedures, Sections 4.2.1, 4.2.2, 4.2.5.

   (d)   Possible deep-seated failure behind tiebacks or beneath the excavation in the case of braced systems - use trial procedures, Section 4.2.1.

   (e)   Possible bearing failure below excavation, Section 4.2.4

   (f)   Possibly instability in soil leading to anchor pull-outs, Sections 4.2.1, 4.2.3.

   (g)   Required wall penetration, Section 4.2.4.

2. **Other Considerations**

   (a)   Influence of tension or shrinkage cracks at top of slope, Section 4.2.3.

   (b)   Effect of seepage forces into excavation, Sections 4.2.1, 4.2.4.

   (c)   Effect of fissuring in stiff clays, Sections 4.2.3, 4.4.1.

4.4   Earth Pressure Loadings for Active or Near-Active Conditions

Tied-back or braced wall systems that are subject to little or no prestress and/or are not exceptionally stiff could develop enough movement to justify use of near-active earth pressure loadings.   Experience and experiment have demonstrated that translation of the wall in the range of 0.1 to 0.25 percent of its height can bring about active conditions. Studies have shown that the total earth loads on a wall can then be predicted using conventional earth pressure theory of Rankine or Coulomb. Allowances may be made in the analyses for curved failure surfaces which yield higher resultant loads than the straight failure surfaces assumed by Coulomb and Rankine. Analyzing for curved failure surfaces, the resultant loading for sands is about 1.1 times that of Coulomb and Rankine and for clays 1.1 to 1.5 times that of Coulomb or Rankine, depending on scale of the excavation and soil strength. In both cases, the resultant is empirically located near mid-height of the wall instead of at the lower third point as in a triangular pressure diagram.

4.4.1 Apparent Loading Diagrams. In design, earth loadings on temporary walls are often determined from empirical loading diagrams rather than from theoretical analyses. Probably the most commonly used are those proposed by Peck, Hanson and Thornburn, Reference 4-14, and shown in Figure 4-5. A number of important points should be remembered concerning these diagrams:

1.   The excavation is assumed to be deep (greater than 20 feet) and relatively wide and wall movements are assumed sufficient to mobilize essentially the full value of soil strength.

2.   Groundwater is assumed to be below the base of cut for sands and for clays, its position is not considered to be influential.

3.   Soil behavior during shearing is assumed drained for sands and undrained for clays; that is, short-term loading conditions only.

4.   Soil masses are assumed homogeneous.

5.   The diagrams are primarily intended for calculating strut loads. Reduction on the order of 33 percent may be appropriate for calculating moments in walls and wales.

6.   Recommended pressures are conservative envelope values intended to account for widely varying observed field behavior and ranges of theoretically predicted behavior.

7.   Loading diagrams apply only to the exposed portion of the walls and not to sections embedded in the ground.

51

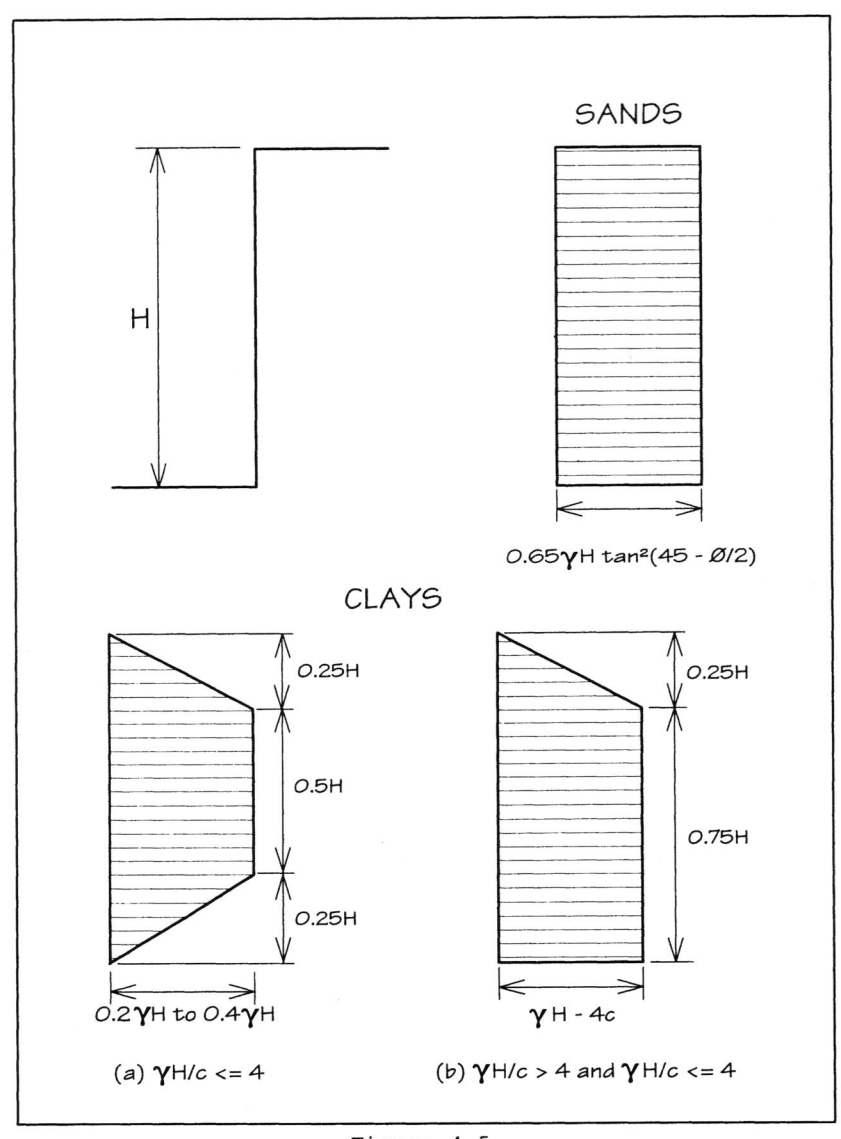

Figure 4-5
Apparent Earth Pressures for Strut Design
(reference 4-14)

As shown in Figure 4-5, the loading diagram for sands is assumed rectangular; the resultant of this diagram is 30 percent greater than the Rankine active resultant for the same ground conditions.

In the case of clays, the level of loading on a support wall is in part related to the degree to which the shear resistance of the soil is mobilized. This is often characterized in terms of the stability number, N, which is defined as $\gamma H/c$. For homogeneous soils, a value of $N = 4$ or less represents the case where a saturated clay can theoretically stand on its own. Thus, in theory, no earth pressure is exerted on a support wall system if N is 4 or less. Should the N value reach 6 or more, the loads begin to increase rapidly since plastic yielding zones are developing in the soil. If the N value is 8 or more and a significant depth of soft clay underlies the excavation, deep-seated soil movements are likely to occur, and loads can be very high.

Two alternative apparent load diagrams are presented by Peck, et al, Reference 4-14, for clays as shown in Figure 4-5, with the recommendation that the one which gives the largest pressures should be used for design. The first diagram has a pressure ordinate defined only in terms of the unit weight of the soil, γ, and the excavation height, H, while the second one uses γ, H and soil cohesion. Typically, the first diagram applies in cases where the N value is 4 or less, a common situation when the clay is relatively stiff or the excavation relatively shallow. Under such circumstances, the loading on a wall system is dictated by factors other than the intact soil strength, and thus, the soil cohesion is not considered in defining the pressure distribution.

If the stability number is greater than four, as for soft clays or deep cuts, then the load diagram (b) of Figure 4-5 is tentatively applicable to design. The resultant of this diagram is about 70 percent greater than the Rankine value. Peck, et al, Reference 4-14, recommends a further check in this case of the basal stability number, $\gamma H/c_b$, where "c_b" is the average cohesion of the clays beneath the bottom of the excavation. If this basal stability number is four or less, then diagram (a) is considered applicable to design.

In instances where stability numbers of 8 to 10 are present and soft clays exist to a considerable depth below the cut, loadings on the wall larger even than those produced by values of diagram (b) have been observed. The large loadings under such circumstances are apparently a result of arching in the soil, caused by large deformations below the excavation. As this occurs, earth pressures on the embedded wall portion are reduced and earth pressures from this wall section are transferred to the upper, exposed wall portion because the wall system here is strutted and thereby much stiffer. These effects are accentuated if the subsoils are strongly anisotropic (Reference 4-4).

53

4.4.2 Alternatives to Apparent Loading Diagram for Effects of Surcharge, Seepage, Nonhomogeneity. The apparent loading diagrams are very useful in bracing design in that they reflect the results of both theory and experience and include a reasonable safety factor for strut design, usually the most critical component of the support system. However, they do not represent the actual pressure distribution on the wall and are cumbersome to use for a markedly stratified profile and to incorporate such effects as seepage, surcharge, sloping soil surfaces or earthquake loadings.

Acceptable alternatives to determine the loading with these complicating conditions are the classical Rankine or Coulomb earth pressure theories. The Coulomb technique, involving trial free bodies in the retained soil, is particularly useful because it can incorporate many complexities which often occur in practice. This method is well described in most geotechnical texts and useful examples are given in Reference 4-12.

Regardless of the analytic method, a theoretical active resultant loading which assumes full soil strength mobilization should be increased by a factor between about 1.1 to 1.5 for design of braces to account for the effects of field variables. In special cases where deep basal heave is a problem, multipliers larger than 1.5 may be justified. The resultant should also be distributed into a rectangular or trapezoidal pressure diagram. An example illustrating the procedure is given in Figure 4-6.

4.4.3 Loading on Embedded Section of the Wall. In most cases of temporary wall design where the base of the excavation is assumed to be stable, loadings on the embedded section of the wall are not considered. It is tacitly assumed that loading on the exposed portion of the wall will govern wall design. However, in very soft clays where basal heave can be a problem, loads on the embedded wall section can be significant, and may, in some particularly critical cases, govern design. In such cases, the embedded wall section must resist horizontal loadings produced by the soil mass which is trying to move into the excavation bottom. The wall section at the lowest brace then in place must be able to accommodate the negative (i.e., cantilever) moment imposed. That brace must resist the horizontal load transferred upward from the embedded section.

Terzaghi, Reference 4-19, proposed a conservative scheme for estimating the load on embedded wall sections in clays where N exceeds 7 which is described in Figure 4-3. The net loading varies depending upon the depth of embedment of the sheetpile relative to the depth of the potential sliding surface. The net load is carried by the wall as in a cantilever about the lowest strut location if the wall only partially penetrates the clay stratum. If the wall fully penetrates to a firm stratum, it can be assumed to act as a simple beam supported at the firm stratum and at the lowest strut location. Depending on specific conditions, some negative moment restraint can be assumed in the wall at the lowest strut point. The critical element

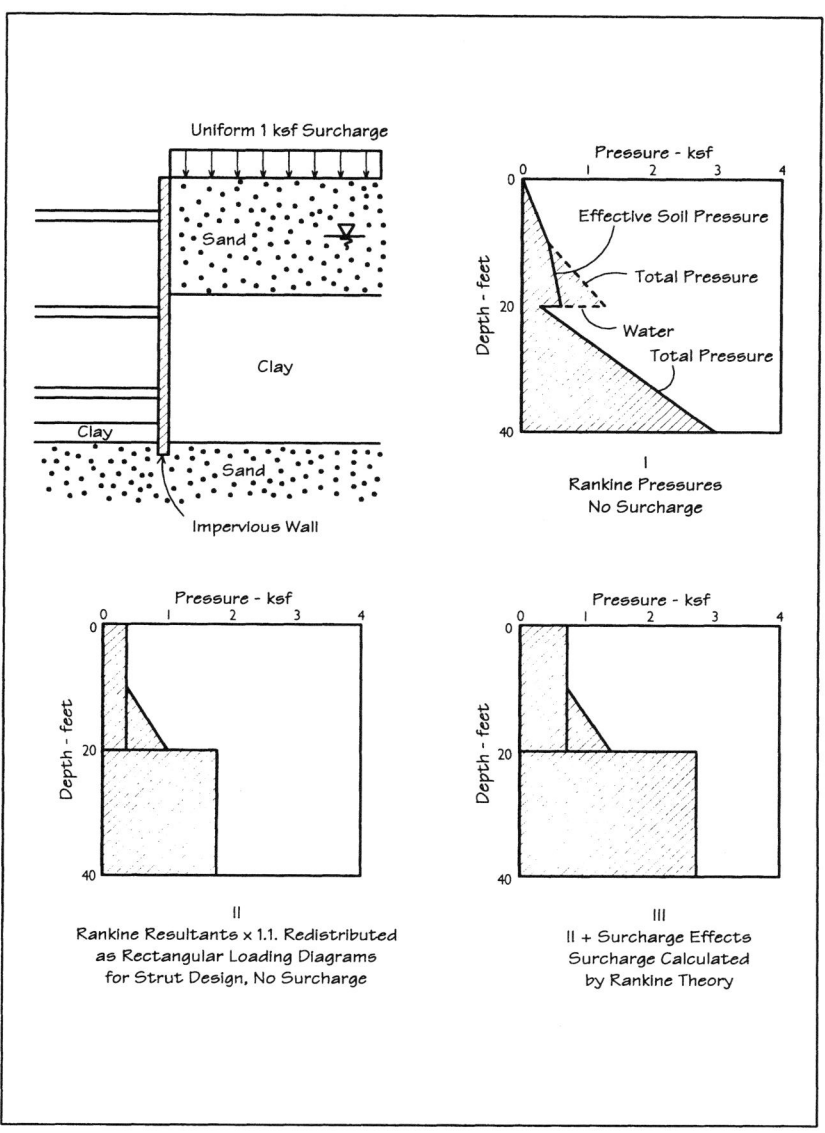

Figure 4-6
Calculating Strut Loading Diagrams
Using Earth Pressure Theory

in th_s case can be the magnitude of positive moment in the wall below the lowest brace.

4.4.4 Charts for Sloping Soil Surfaces and Sloping Walls. The case of a combination of sloped cut and supported wall for an excavation is relatively common. A sloped, braced or tied-back wall is used on some occasions. In either case reasonable estimates for the earth loading can be determined from charts using Coulomb or Rankine procedures for active loadings produced by cohesionless soils as shown in Figure 4-7. In the general case, the resultant loading from the char_ should be increased by a factor of 1.1 to 1.5 for design of struts.

4.4.5 Increased Earth Loadings Due to Earthquake. Earthquake loadings are generally not considered for temporary wall systems. However, in an area of high seisnic risk and substantial time exposure, it may be necessary to account for the poss bility. Liquefaction which can apply a horizontal pressure equal to full vertical overburden weight on the wall, is covered in Section 4.7.3. For systems where liquefaction is not a problem, the Coulomb analysis has been adapted for earthquake loading effects caused by ground shaking (Reference 4-17). The increased earth loading is a function of soil strength and ground acceleration. A char_ is provided in Figure 4-8 to facilitate evaluation of the extra earth loading due to ground acceleration. The dynamic earth loading is applied at a point one-third of the wall height down from the ground surface.

4.5   Earth Pressure Loadings for Other than Active Conditions

4.5.1 Criteria for Greater than Active Loadings. If the wall system is highly preloaded and/or stiff and workmanship is of high quality, movements are minimized. In this case it is possible that the wall will be acted upon by loadings larger than those predicted by methods covered in the preceding sections.

In o-der for a wall system to be loaded by greater than active loadings, one of two criteria must be met:

1.   The wall system must be so stiff that lateral movements will be less than 0.1 percent of the wall height. Few systems ever practically achieve this standard of performance without some preload since, even if the wall is stiff, the movements are also influenced by brace compression or tieback extension and workmanship. Evidence is clear that most wall systems which are not prestressed deflect the small amount required to develop active loadings.

2.   The total preload must be greater than the active resultant and the moment of preload about the center of potential soil mass rotation must exceed that

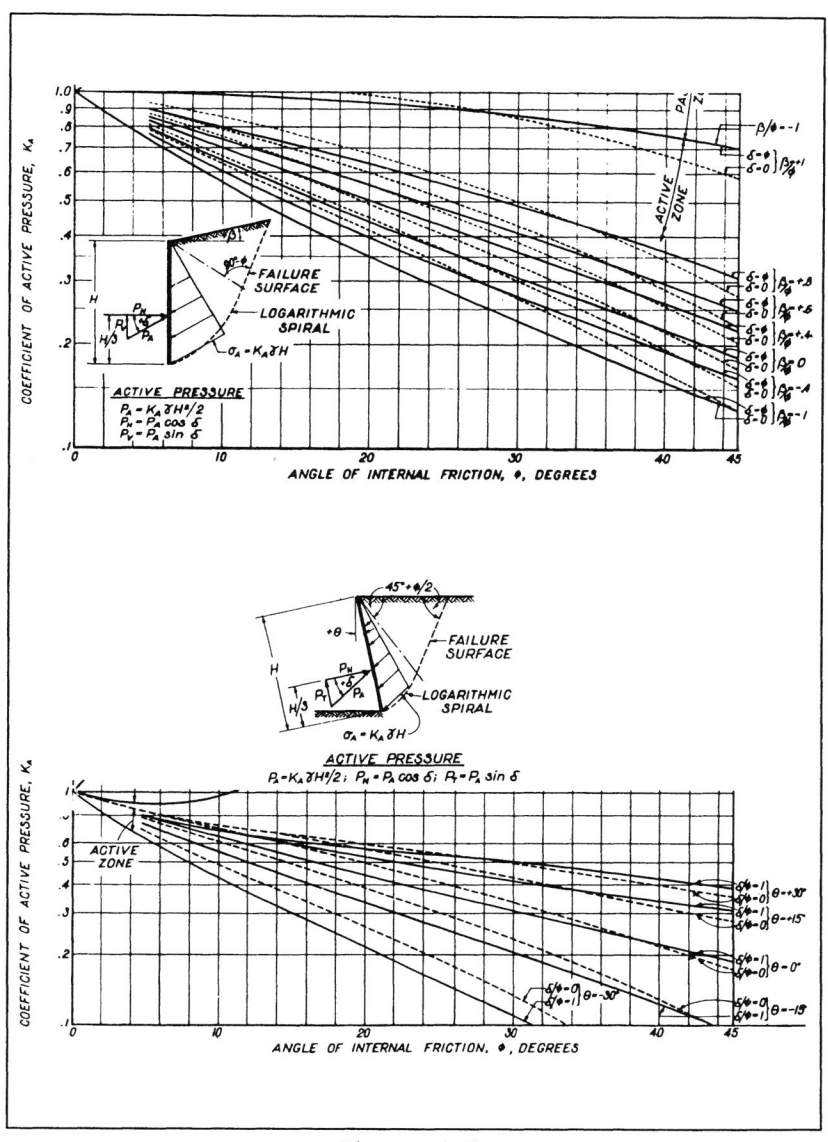

Figure 4-7
Active Earth Pressure Coefficients
for Sloping Walls and Soil Surfaces
Cohesionless Soils - Wall Friction = 0
(reference 4-12)

57

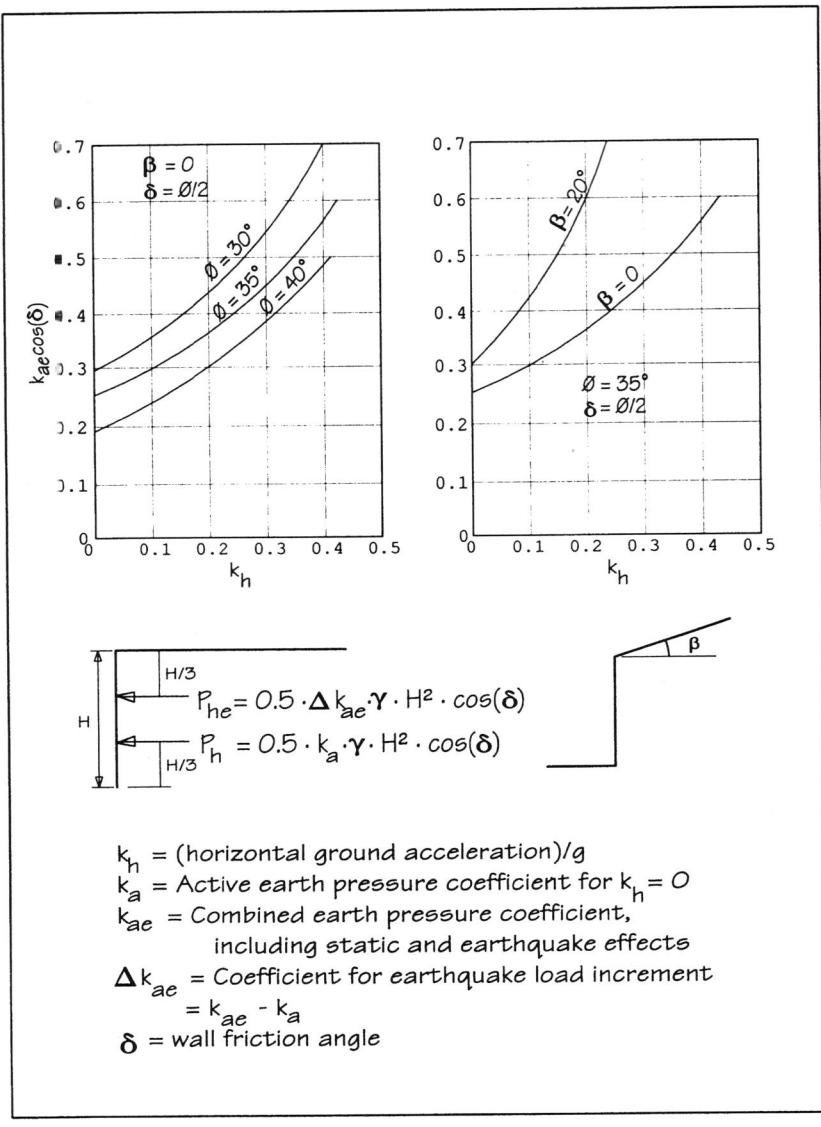

Figure 4-8
Additional Loading Due to Earthquake
for Walls in Cohesionless Soils
(reference 4-17)

of the active resultant. Preloads must be stable and not subject to reduction by creep of anchors or bracing.

The second criterion can be met relatively easily for systems where the soil mass beneath the excavation is not subject to base failure.

4.5.2 Design Loadings. For a wall that is not preloaded but is perfectly stiff and immovable, earth pressure loading should be that of the at-rest condition. Since no temporary wall systems are perfectly stiff, the earth loading for a stiff system should lie somewhere between active and at-rest resultants and it is probably justifiable to design for loadings nearer active than at-rest.

For heavily preloaded walls, as defined by the second criterion in Section 4.5.1, the total earth pressure resultant should equal the total horizontal preload. Distribution of the loads is affected by soil stiffness, wall stiffness and preload spacing. For a refined evaluation of these effects, analytical approaches such as the beam on elastic foundation method or the finite element method may be employed. However, for approximate design purposes, the apparent pressure diagram (horizontal load divided by spacing between preloads) may be used.

Because calculating the actual loadings on a wall subjected to greater than active pressures is an indeterminate problem, analysis of surcharging, sloping backfills, etc., is not strictly amenable to the approaches described in Section 4.3. However, a Coulomb-type analysis may be used to estimate approximate values for the effects of these extra loadings.

4.6   Effects of Groundwater Loadings

Water may impose loads on the wall system if the water table is not drawn down below the excavation base. A detailed analysis of this condition requires knowledge of soil permeability, possible wall leakage, flow patterns, and groundwater conditions. Water loads are added to effective earth loads obtained by the procedures described in earlier sections of this chapter.

Two conditions of water loading are shown in Figure 4-9. In the top of the figure, an impervious wall penetrates to an impervious layer, and the wall is loaded by a hydrostatic water pressure distribution. Seepage, either through the wall at faulty joints or beneath the wall reduces water loading. Effects of seepage through a pervious soil are shown in the second diagram in Figure 4-9 where it can be seen that water loading on the wall is reduced from a hydrostatic condition. Seepage into the excavation can, however, produce a downdrag loading on the wall. In this case, effective stresses in the soil are increased and settlements are produced in the soil retained behind the wall.

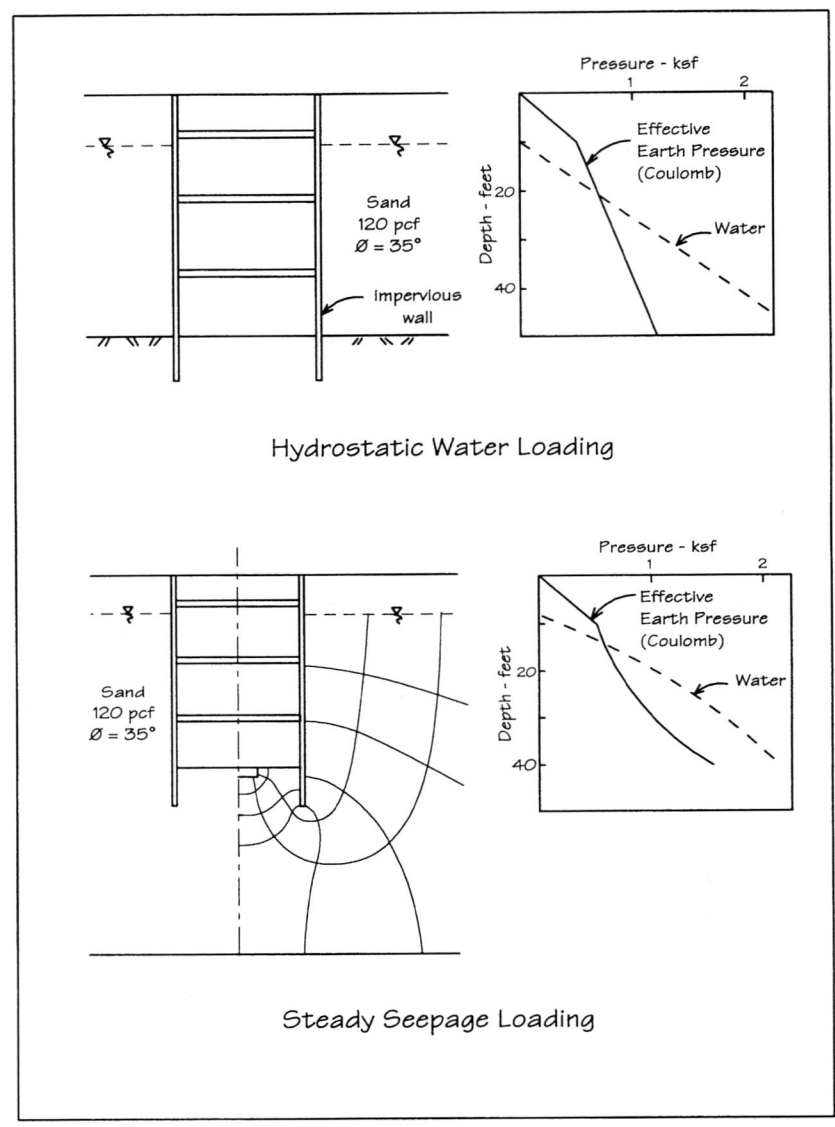

Hydrostatic Water Loading

Steady Seepage Loading

Figure 4-9
Earth and Groundwater Loadings
for Problems with Groundwater Table

60

From the examples in Figure 4-9, it is apparent that water loadings on impervious walls can be even more important than earth loadings. On the active side of the wall, seepage reduces water pressures and at the same time increases the effective earth pressures due to the downward seepage load. However, on the passive side of the wall, the opposite behavior occurs. The earth pressure increases are more than offset by decreases in water loading. Effects of seepage on effective soil pressure may be calculated by Coulomb earth pressure theory, Reference 4-12. This utilizes boundary water pressures determined from a flow net for the free body of retained soil bounded by straight failure planes.

Unfortunately, the extent of the seepage effect cannot be precisely evaluated in many cases. For example, seepage in fine grained soils may not develop depending upon the time the excavation is open and recharge conditions. The chances of seepage developing through the wall are a function of the type of wall, construction procedures, and soil conditions. Steel sheetpile walls can range from almost completely pervious to impervious depending on the tightness of interlocks between adjacent sheets and the relation of wall permeability to soil permeability. For design, a reasonable range of permeabilities should be considered.

4.7   Ancillary Loadings

In addition to the loadings described above, extra loadings on braced systems can be caused by frost action, temperature fluctuations on braces and liquefaction of soils during an earthquake.

4.7.1 Frost Action. Periods of cold temperature can cause frost penetration of the soil and result in significant expansion and movements of the soil which increase bracing loads. Measurements of strut loads in braced excavations in Norway during the winter months have shown that strut loads are, in some instances, tripled by frost penetration effects. Measured loads on tiebacks in Sweden have also shown large increases due to frost action, in some cases leading to tie failure. In areas where the potential for frost action exists, careful consideration should be given to such effects.

4.7.2 Temperature Fluctuations on Braces. Braces are often exposed to daily and seasonal fluctuations of temperatures. These fluctuations result in a lengthening and shortening of the braces and related fluctuations in brace loads. Observed load changes in struts due to temperature are in general not large enough to cause failure, but can lead to loss of prestress or be a particular nuisance in interpretation of instrumentation data.

For example, a steel strut with an area of 30 square inches and subjected to a temperature increase of 40 degrees F would increase its load by 90 tons, a stress increment of 6 ksi, under complete end restraint. Fortunately, only a small change in length is required to relieve the load change and unless the ends of the strut are truly immovable, the actual load increase will be much smaller. For each job, the circumstances should be evaluated as to potential temperature changes and end conditions of the struts. Allowances should be made in selecting design working stress for some increase or decrease in strut load under temperature fluctuation.

4.7.3 Earthquake Liquefaction of Saturated Sands. During an earthquake, loose to medium dense saturated sands and silts have the potential to liquefy or lose most of their ability to resist shear stresses. Generally, such problems are not considered in design of temporary retaining structures because of the small likelihood a seismic event large enough to cause liquefaction will occur during the working life of that structure. If local codes require this to be considered, Seed, Reference 4-16, has proposed a useful and relatively simple procedure for evaluating this potential. If the supporting soil at the interior subgrade can liquefy, little can be done to maintain stability of the wall unless the wall can be keyed into a more earthquake resistant stratum below the vulnerable soil. If the soil being retained is vulnerable to liquefaction in a seismically active area during the life time of the system, then the wall should be designed for horizontal pressure equal to full vertical overburden weight.

4.8 Design Checklist for Loading of Supported Excavations

In the preceding sections, a variety of loading conditions and methods for calculating loads on supported excavations have been discussed. In this section, a design checklist is provided as a quick reference for the designer.

Loading Checklist

1. **Basic Problems**

    (a) Effect of system stiffness and degree of prestressing, Sections 4.4.1, 4.5.1, 4.5.2.

    (b) Basic loading of wall systems, Sections 4.4.2, 4.4.3, 4.5.2.

    (c) Effect of deep-seated movements, Section 4.4.1, 4.4.3, 4.2.4.

    (d) Loading on embedded portions of walls, Section 4.4.3.

    (e) Effects of groundwater loadings, Section 4.6.

2. **Other Considerations**

    (a) Surcharge loadings, Section 4.4.2.

(b)   Effect of sloping soil surfaces, Section 4.4.4.

(c)   Effect of wall slope, Sections 4.4.4.

(d)   Earthquake loadings, Sections 4.4.5, 4.7.3.

(e)   Loadings due to frost action, Section 4.7.1.

(f)   Loadings caused by temperature fluctuations, Section 4.7.2.

4.9   Selection of Shear Strength Parameters

In Chapter 2, types of shear strengths and methods for soil sampling and strength testing were discussed. In the evaluation of stability and earth loadings, the shear strength of the soil is a fundamental parameter. The primary problem in selection of shear strength arises for clayey and silty soils when undrained conditions apply. The selection process cannot be rigorously codified since the parameters one would use for design will depend not only upon the basic results, but also upon homogeneity and sensitivity of the clay, the level of experience in a certain environment, the degree of stability of the cut, and the exposure of the site. For example, the process of selection of strengths for a stiff fissured clay or a highly sensitive, soft clay will be different. If safety against basal heave of a cut is very low, more conservatism and consideration of effects such as creep and anisotropy need to be given than if the cut is very safe against basal heave. The following examples are provided to deal with typical situations where unusual effects do not have to be considered.

4.9.1 Soil Strength Increases with Depth. In Figure 4-10, an undrained strength profile is shown for a clay with a thin overconsolidation crust above essentially normally consolidated clay. Laboratory test strength is high in the overconsolidated crust, becomes lower below the crust and then increases with depth. Selection of design shear strengths for this clay deposit depends upon the depth of wall and the type of analyses. Two examples are illustrated in Figure 4-10 for a relatively shallow and a relatively deep cut. In either case, strength assigned to the stiff crust should be selected conservatively. This crust is likely to be naturally fissured and subsequently cracked either by horizontal extension strains or large movements in the underlying soil. Strength increases in the lower clay may be accounted for either by using average undrained shear strength values or by considering the lower clay segmented in a series of strata with successively larger average strength values. The latter approach is recommended for analyses which can accommodate layering. To utilize the empirical apparent pressure diagrams, overall average values must be used. The effect of cut depth on average value used is shown in Figure 4-10. For the deep excavation, the value of undrained shear strength is about twice that used

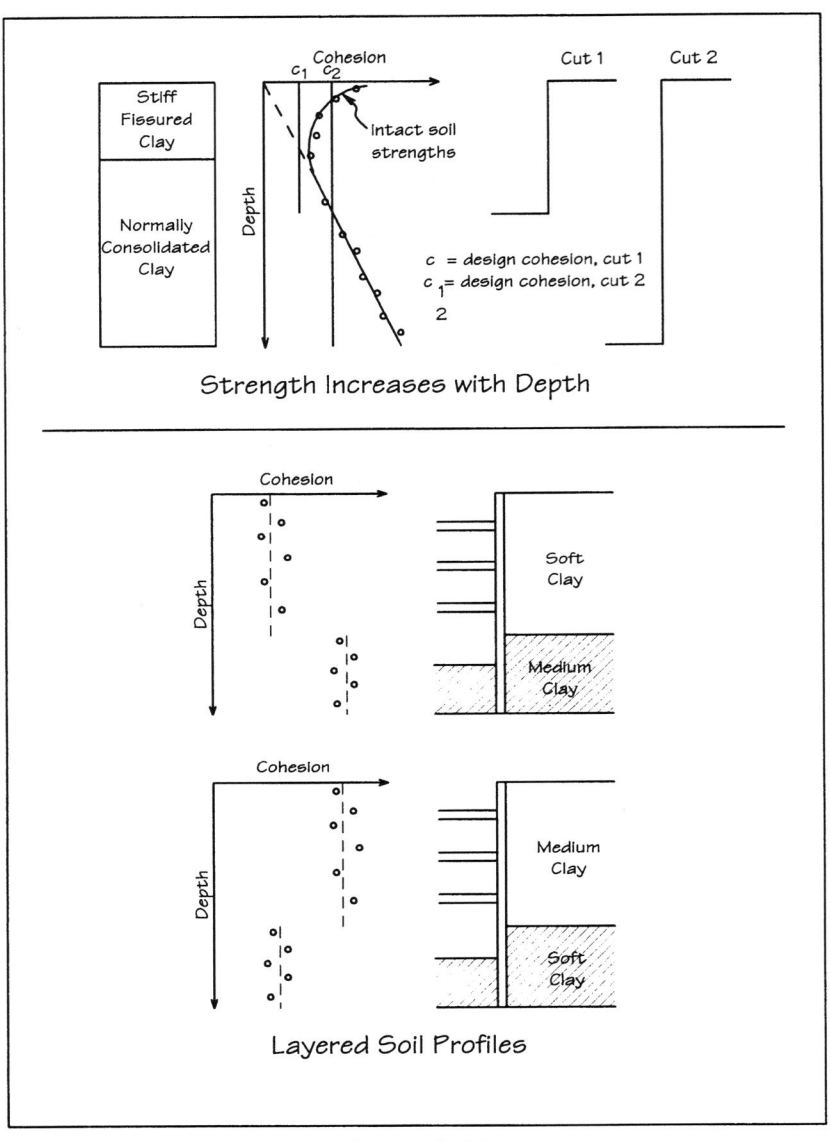

Strength Increases with Depth

Layered Soil Profiles

Figure 4-10
Selection of Strength Values for Analysis

64

for the shallow cut. Stability against basal heave and the stability number, $\gamma H/c$, should be evaluated more in terms of the strength near the bottom of the cut rather than average undrained strength.

4.9.2 Layered Soil Profile, Soft to Medium Clays. Two cases involving layered soil profiles are shown in Figure 4-10. In the first instance, a soft clay overlies a medium clay and, in the second, a medium clay overlies a soft clay. Strengths for design will depend upon the type of analysis performed and the threat of basal failure. If trial analyses are used, then specific variations of the strengths may be used. If empirical apparent loading diagrams are used, the following guidelines are recommended:

1. Evaluate basal failure potential and stability number, $\gamma H/c$, in terms of strength of the underlying soil stratum.

2. If there is a strong potential for basal failure and in the case where the soft clay is the lower layer (a common condition), then the wall loading should be evaluated in terms of strength values of both layers with an emphasis on strength of the lower.

3. If the factor of safety against basal failure is adequate, then the loading on the wall should be evaluated in terms of the weighted average strength of the retained soils.

4.10   Safety Factors Incorporated into Design Procedure

Because of the large element of judgement involved in the analysis and design of a temporary support system, it is relatively easy to incorporate either too large or too small a factor of safety. In Chapters 5, 6 and 7, guidance is given on safety factors or stress levels to be used in component design of the structural elements. It is important to realize, however, that hidden effects on the overall factor of safety of the system can be introduced in: (1) the procedures employed to sample and test the soils for strength parameters; (2) the methods used to interpret the soil test data; (3) the types of loading and stability analyses used; and, (4) the additional loads added in the analyses to account for surcharging, earthquake, temperature, etc. The effects of these hidden factors are often cumulative, although not necessarily conservative.

For example, in an extreme case, a designer may rely on a conservative load analysis procedure coupled with inferior sampling and testing techniques which typically lead to soil disturbance and lowered laboratory strengths, conservative evaluations of the poor laboratory data and overly safe estimates of surcharge loads. Then, in the final design, the designer incorporates conventional structural safety factors and concludes that the support system has the safety margin of the structural analysis. In fact, the system is likely to be over-designed. In performing analysis and design of temporary walls, each step of the work must be considered for the factor of safety introduced. A dangerous possibility is that some aspect of the ground conditions is overlooked, such as base instability, the threat of piping from below or the presence of obstructions. One of the most effective ways of guarding against this is to study local experiences with braced excavations.

4.11 References

4-1   Bishop, A. W. (1955). "The Use of the Slip Circle in the Stability Analysis of Slopes," *Geotechnique*, 5(1), 7-17.

4-2   Bishop, A. W. and Morgenstern, M. (1960). "Stability Coefficients for Earth Slopes," *Geotechnique*, 10(4), 129-150.

4-3   Clough, G. W. and Denby, G. M. (1977). "Stability Berm Design for Temporary Walls in Clay," *Journal of the Geotechnical Engineering Division*, ASCE, 103(GT2), 75-90.

4-4   Clough, G. W. and Hansen, L. A. (1981). "Clay Anisotropy and Braced Wall Behavior," *Journal of the Geotechnical Engineering Division*, ASCE, 107(GT7), 893-914.

4-5   Hoek, E. and Bray, J. W. (1973). *Rock Slope Engineering*. Institute of Mining and Metallurgy, London.

4-6   Janbu, N. (1954). "Stability Analysis of Slopes with Dimensionless Parameters," *Harvard Soil Mechanics Series No. 46.*

4-7   Janbu, N. (1957). "Earth Pressures and Bearing Capacity Calculations by Generalized Procedure of Slices," *Proceedings*, 4th International Conference on Soil Mechanics and Foundation Engineering., Vol. 2, 207-212.

4-8   Jumikis, A. R. (1962). "Active and Passive Earth Pressure Coefficient Tables," *Engineering Research Publication No. 43*, Rutgers University, NJ.

4-9   Kenney, T. C. (1963). "Stability of Cuts in Soft Soils," *Journal of the Soil Mechanics and Foundations Division*, ASCE, 89(SM5), 17-37.

4-1C   Morgenstern, N. R. and Price, V. E. (1965). "The Analysis of the Stability of General Slip Surfaces," *Geotechnique*, 15(1), 79-93.

4-11   National Research Council of Canada (1975). *Canadian Manual on Foundation Engineering*. National Research Council of Canada, Ottawa.

4-12   Naval Facitilities Engineering Command (1986). "Soil Mechanics: Design Manual 7.1," *NAVFAC DM-7.1*, Department of the Navy, Alexandria, VA.

4-13   O'Rourke, T. D., Cording, E. J., and Boscardin, M. (1976). "The Ground Movements Related to Braced Excavations and Their Influence on Adjacent Building," *Report No. DOT-TST 76T-23*, U. S. Department of Transportation.

4-14   Peck, R. B., Hanson, W. E., and Thornburn, T. H. (1974). *Foundation Engineering*. 2nd Ed., John Wiley and Sons, 1974.

4-15   Ranke, A. and Ostermayer, H. (1968). "Beitrag zur Stabilitaetsuntersuchung mehrfach verankerter Baugrubenuschliessungen (A Contribution to the Stability Calculations of Multiple Tied-Back walls)," *Die Bautechnik*, 45(10), 341-349.

4-16   Seed, H. B. (1979). "Soil Liquefaction and Cyclic Mobility Evaluation for Level Ground During Earthquakes," *Journal of the Geotechnical Engineering Division*, ASCE, 105(GT2), 201-256.

4-17   Seed, H. B. and Whitman, R. V. (1970). "Design of Earth Retaining Structures for Dynamic Loads," *Proceedings*, ASCE Specialty Conference on Lateral Stresses and Design of Earth-Retaining Structures, Cornell University, Ithaca, NY, 103-149.

4-18   Taylor, D. W. (1948). *Fundamentals of Soil Mechanics*. John Wiley & Sons, New York, NY.

4-19   Terzaghi, K. (1943). *Theoretical Soil Mechanics*. John Wiley & Sons, New York, NY.

CHAPTER 5: WALLS OF BRACED AND TIED-BACK EXCAVATIONS

5.1 Introduction

The wall of an excavation support system comprises those structural members in direct contact with the retained soil. Resistance to ground movement provided by the system largely depends on the performance of the bracing or tiebacks which support the wall, but also is influenced by the tightness of contact between the wall and support elements, flexural stiffness of the wall, and the wall's ability to resist deep-seated movement. The wall serves the following general functions:

1. It has sufficient tightness or continuity to avoid sloughing or ravelling of the retained soil. Where stiff cohesive soils, cemented soils, or weak rock is retained, a wall element may not be required. In these cases, localized use of shotcrete or other soil stabilization techniques and materials may be sufficient to provide the required support.

2. In some situations it is intended to allow groundwater to drain through openings without erosion. In others, the wall prevents flow of water into the excavation.

3. The wall may penetrate deeply enough below the excavation to provide diversion of seepage paths to resist piping and heave in the excavation bottom.

4. The wall resists bending stresses between supports and below the excavation base. It transfers loads at connections and must limit bending deflections to tolerable amounts between braces and below the lowest brace. The embedded bottom of the wall may support vertical loads as well as lateral loads.

5. The wall often serves as a form for the foundation wall of the permanent structure. It can be used to hold forms and waterproofing for the structure. In special cases the wall may serve as the permanent exterior of the structure foundation.

6. The wall must be appropriate for the particular subsurface conditions so that it can act with the retained soil and other components of the retention system to provide stability. The wall section chosen must be able to be installed at the site on the intended alignment without damage and within certain geometric tolerances.

Walls may be constructed of rolled or fabricated steel sections, precast-prestressed concrete, cast-in-place reinforced or unreinforced concrete, wood or combinations

68

of these materials. A variety of construction methods are appropriate. Walls can be completely or partially formed by driven elements. Wall panels may be created by excavating the soil and filling the trench with cast-in-place concrete or by placing precast concrete sections. Each method and wall type has distinctive performance and construction characteristics which influence its selection. These characteristics include stiffness, ease of handling and installation, durability, watertightness or continuity, and ease of removal. Table 5-1 summarizes these properties in brief. Watertightness and permeability characteristics are listed in Table 3-1.

The principal wall types are described in detail in the following subsections. They consist of four major categories: driven sheetpiling, soldier pile and lagging walls, various types of cylinder walls, and concrete diaphragm or slurry walls. Some less conventional wall types are also noted. Reference specifications for wall materials are given in Table 5-2.

5.2   Driven Sheetpile Walls

A common wall type consists of long, relatively slender sheets of wood, steel or concrete driven into the ground before excavation. A continuous soil-tight and watertight membrane is often intended but not always achieved. The resulting wall is moderately flexible and horizontal movements due to bending of the wall may be higher than for the other principal types. Steel sheetpiles are fabricated with interlocking edges. Wood and concrete sheetpiling often employ a tongue and groove edge arrangement for continuity. For deep excavations the principal sheetpile material is the steel Z-shaped section.

Steel sheetpiles are typically used in potentially squeezing or running soils such as soft clays and cohesionless silt or loose sand below the water table. These materials can be unstable when exposed during excavation. Placement of lagged walls against them can be difficult, if not impracticable. Interlocking steel sheetpiling can intercept concentrated seepage through pervious layers and counter the possibility of piping from below the excavation. When an excavation is dewatered, a steel sheetpiling wall does not necessarily prevent lowering of piezometric levels in the retained soil since the wall interlocks can allow some leakage. Consolidation and settlement of adjacent areas could therefore occur when the soils surrounding the excavation are compressible. In compact to very compact granular soils that contain cobbles and boulders or where the rock surface is above excavation base, sheetpiling is not normally used.

5.2.1 Section Properties.   Detailed information on steel sheetpile sections is provided in the catalogs of principal manufacturers. The conventional ASTM grade used for American sheetpiling is A328 which has a minimum yield point of 38.5 ksi (266 MPa) and a normal AISC working stress of 25 ksi (173 MPa). In

69

TABLE 5-1. Properties of Braced and Tied-Back Walls

SYSTEM	PROPERTIES EI kst/ft x 10³	MOMENT kip-ft/ft	DEPTH RANGE feet	WATER-TIGHTNESS	ADVANTAGES	DISADVANTAGES
t = 2" to 4" VERTICAL WOOD SHEETING	0.1 to 1.0	1 to 3	5 to 20	No	Lowest cost, ease of installation in good ground. Uses conventional equipment and simple skills	Discontinuous, low strength, limited soil conditions, limited depth
VERTICAL STEEL SHEETING	3 to 50	10 to 125	15 to 70	Fair	Continuous, high strength, readily available, effective in soft ground.	Limited by soil conditions, hampered by obstructions
t = 3" to 6" VERTICAL PRECAST CONCRETE SHEETING	1.5 to 10	2 to 10	10 to 30	Fair	Durable, cost effective, resists soil piping.	Limited availability, can be damaged in handling and driving, limited depth
8WF to 14WF @ 6' to 8' CC SOLDIER BEAM & WOOD LAGGING	3 to 40	7 to 70	15 to 60	No	Ease of installation in good ground, readily available, adaptable to poor ground, low cost.	Poor tolerances in hard ground, loss of ground common, may need precored holes
8WF to 14WF @ 6' to 8' CC 4" to 6" SOLDIER BEAM & CONCRETE LAGGING	3 to 40	7 to 70	15 to 60	Fair	Adaptable to poor ground, could use spray-on coating of shotcrete.	Same problems with soldier pile driving, face in soil must remain open for several feet height

TABLE 5-1. Properties of Braced and Tied-Back Walls (cont.)

SYSTEM	PROPERTIES				WATER-TIGHTNESS	ADVANTAGES	DISADVANTAGES
	EI	MOMENT	DEPTH RANGE feet				
	ksf/ft x 10³	kip-ft/ft					
STEEL BEAM OR BAR REINFORCED d=18" to 36" CYLINDER PILES - TANGENT	70 to 600	90 to 350	30 to 60	Poor	Common technique, piles can be widely spaced in good ground & can be made as stiff as desired by adding core	Poor tolerances in hard ground. Potential difficulty with soil piping.
18" to 36" DIAMETER CYLINDER PILES - SECANT	80 to 700	100 to 400	20 to 40	Fair	Improved water-tightness, piping more readily controllable. Adaptable to irregular layout.	Poor tolerances in hard ground. Difficult to maintain overlap.
CYLINDER PILES - STAGGERED	100 to 850	125 to 500	20 to 40	Fair	Common technique. Core can be added as needed for stiffness. All cylinder pile walls give measure of protection to nearby structures.	Poor tolerances in hard ground.
t = 24" to 36" SLURRY WALL - STEEL BEAM REINFORCED	350 to 1600	30 to 400	20 to 100	Good	High strength, durable. Can eliminate underpinning. Can be permanent wall. Good tolerances.	Higher cost. Special equipment. Stringent quality controls
t = 24" to 36" SLURRY WALL - REINFORCING BARS	300 to 1000	15 to 260	20 to 100	Good	High strength, durable. Can eliminate underpinning. Can be permanent wall. Good tolerances.	Higher cost. Special equipment. Stringent quality controls.

Note: Moment is given as typical range of values that can be accommodated by the wall at normal working stress, averaged per lineal foot of the wall.

Table 5-2. Reference Standards and Specifications

MATERIAL	USE	REFERENCE
Lumber	Trench bracing	For current summary of grading classification, properties, quality control see NBS Building Science Series 122, "A Study of Lumber Used for Bracing Trenches in the United States."
Steel Structural Elements	Soldier piles, wales, braces, struts, connections	AISC, "Manual of Steel Construction"
	Structural sections	ASTM A36 and A572
	Steel Sheet Piling	ASTM A328, A572, and A690
	Pipe piles and struts	ASTM A252
	Welding	AWS-W1-1
Concrete Structural Elements	Cylinder piles	ACI 318, "Building Code Requirements for Reinforced Concrete"
	Concrete lagging	ACI 336, "Standard Specifications for End Bearing Drilled Piers"
	Precast concrete sheeting	ACI 543, "Recommendation for Design, Manufacture and Installation of Concrete Piles"
	Diaphragm walls	ACI 336, "Suggested Design and Construction Procedures for Pier Foundations"; Federation of Piling Specialists, "Specification for Cast in Place Concrete Diaphragm Walling".

72

Table 5-2. Reference Standards and Specifications (cont.)

MATERIAL	USE	REFERENCE
Components: Cement, Aggregates, Reinforcing Steel, Steel Tendons	Structural concrete	ACI 322
	Precast concrete	ACI 512
	Portland cement	ASTM C150
	Concrete aggregates	ASTM C33
	Reinforcing bars	ASTM A615
	Wire mesh	ASTM A185
	Tieback anchors	"Recommendations for Prestressed Rock & Soil Anchors" (PTI)
	Strand or cable	ASTM A416
	Wire	ASTM A421
	High strength bars	ASTM A322
Bentonite	Stabilizing fluid for diaphragm wall construction	API Spec. 13A, "Oil Well Drilling Fluid Materials" API RP 13B, "Standard Procedures for Testing Drilling Fluids"

NOTES AND ABBREVIATIONS:
ACI - American Concrete Institute
API - American Petroleum Institute
AWS - American Welding Society
PTI - Post-Tensioning Institute
AISC - American Institute of Steel Construction
ASTM - American Society for Testing and Materials

American practice it is assumed that shear resistance cannot develop along the interlocks and therefore the section modulus of deep-arch and medium-arch sheets, which are interlocked on the walls' midplane, is considered to be that of a single sheetpiling. The section modulus cannot be combined with that of adjacent sheetpiles to provide for resistance to bending. Some European and Asian practices differ in this respect, in that they assume interlock friction is effective, and thus take advantage of the full section modulus of adjacent sheets acting together.

5.2.2 Installation Problems. Installation of steel sheetpiling by driving follows a well-established sequence with the tongue or ball end always leading and the vertical offset between adjacent sheetpiles limited to no more than about five feet. At many sites it is necessary to excavate a trench before driving piles in order to clear obstructions down to the lowest level of previously existing structures. If obstructions are encountered during installation which cause sheetpiles to jump out of interlock, driving should be discontinued until the obstruction can be cleared by direct excavation, or is broken or bypassed by spudding or jetting. If this is not practical and driving is continued, supplementary lagging or additional sheeting may be needed to repair the damage. A vibratory hammer can aid in penetration of compact granular material. The effects of vibratory sheet pile driving on surrounding ground are discussed in Chapter 8.

5.2.3 Extraction of Sheeting. Extraction of sheeting can be facilitated by use of vibratory extractors. Removal in cohesive soils, even with a high speed hammer, may leave an open slot by withdrawing adhering soil. This slot can contribute to displacement of the surrounding ground. If sheeting is to be left in place, it is usually desirable to cut it off at a level no higher than 3 to 6 feet (1 to 2m) below the final ground surface to avoid forming a hard point which can result after eventual ground adjustments.

5.3 Soldier Pile and Lagging Walls

Soldier piles are vertical steel or concrete structural sections installed at horizontal spacing between about 5 and 8 feet (1.5 and 2.4m). Timber lagging is placed horizontally to retain the earth between piles as excavation proceeds. When lagging is used, the soil must have enough cohesion to remain stable after trimming and before the lagging is inserted. Some stiff cohesive as well as cemented soils permit construction of soldier pile walls without lagging, but even in these cases, a binding coat of shotcrete is sometimes sprayed on the soil surface to resist ravelling.

Soldier pile and lagging walls are probably the most widely used form of ground support. They have the advantage that steel soldier piles can be driven through compact or irregular strata that would obstruct sheetpiling. When driving is unacceptable or impractical, holes to receive soldier piles can be predrilled through soils or soft rock. Predrilling in granular or silty soil below the water table should be performed under a full head of drilling fluid to stabilize the hole and prevent local collapse and sloughing which can cause adjacent ground movements. When the soldier pile is installed, the hole is usually filled with tremied lean concrete or sand-cement grout. Soldier pile installations are adaptable to existing underground utilities or structures which can be avoided by judicious layout of the piles; lagging can be worked around or beneath an obstruction which must remain in place. The system can also be adapted to an irregular wall alignment. A properly lagged wall permits drainage, drawdown and reduction of exterior water pressures without flow of retained soil.

5.3.1 Soldier Piles. The most common soldier piles are rolled steel wide-flange or bearing pile steel sections, but they can be almost any structural member, including timber, fabricated steel pipe sections, cast-in-place concrete or precast concrete elements. A bearing pile section often will be used when soldier piles are expected to encounter hard driving because of its resistance to twisting and distortion. Deeper wide flange sections are used when greater bending stiffness is required to resist lateral deflections. These may preferably be set in pre-excavated holes. A variation of this is a patented procedure wherein soldier piles are driven on a slight batter for the purpose of decreasing the weight of soil in the active zone and consequently reducing the lateral force on the support system.

Built-up structural steel sections have been utilized for special purposes. Back-to-back channels may be placed in pre-excavated holes and tiebacks installed in the slot between channels. Pipe piles which have been installed for underpinning can be utilized as soldiers by welding or bolting a T-section to the front of the pipe to retain wood lagging. Steel plate or T-sections also can be welded to wide flange beams to provide greater bending resistance. For cast-in-place concrete soldier piles a hole is pre-excavated, the reinforcing cage lowered and concrete placed with a foam plastic insert on the line of the wall, which, when removed from the concrete, provides a recessed seat for lagging. Precast concrete H-piles have also been utilized, and may be either driven or lowered in pre-excavated holes.

5.3.2 Lagging Materials. Lagging between soldier piles is most commonly wood but can also consist of light steel sheeting, corrugated metal, precast concrete, or pneumatically applied cement-concrete. Where rock or hard cohesive soils are retained, the material exposed in the space between soldier piles may simply receive a spray-on coating of asphalt or shotcrete. Wire mesh and rock bolts may be placed to prevent fall of fragments. Wood lagging may be placed either behind or in front of the flange of a steel soldier pile next to the excavation. For other pile types a channel or T-section may be attached to hold lagging. There are several examples of the use of pneumatically-applied cement-concrete over wire mesh between concrete soldier piles, or of concrete walls cast between steel soldier piles. In the latter case a face roughly three to five feet in height must be opened

75

without support for casting the concrete panel. Generally an equivalent exposure is necessary for the use of pneumatically applied cement-concrete.

5.3.3 Selection of Wood Lagging. By far the most common American procedure is to utilize wood lagging. Generally a species is selected with an allowable flexural stress in excess of 1000 psi (6.9 MPa). Often lagging from previous jobs is used. In this case it should be sound and undamaged. Because of the well-known effect of arching of earth pressures away from the midpoint of the lagging to the more resistant soldier piles, the design of lagging to withstand uniformly-distributed earth pressures usually is unrealistic. Lagging thickness is selected on the basis of empirical rules and experience which depend on the character of retained soil, spacing of soldier piles and depth of excavation. In ordinary circumstances with usual soldier pile spacings, 3 inch (75mm) thick rough wood lagging is used down to depths of 15 to 30 feet (5 to 10m) and 4 inch (100mm) wood is used below this level. With more widely spaced soldier piles [up to 10 feet (3m)] or in soft to medium clays or uniform sands below the water table, lagging thicknesses may be increased. Fractured or slickensided, heavily overconsolidated clays may appear deceptively strong in the subsurface investigation. However, high residual stresses, decrease in strength due to opening of fractures, or movement on the slickensides can produce bulging of lagging and other evidence of high lateral pressures and make such clays difficult and treacherous to retain.

5.3.4 Installation Problems. Installation problems may be grouped according to control of ground water or surface water, installation of soldier piles, or placing of lagging.

1.   Control of surface and ground water flow through a soldier pile wall can be critical to its success. Surface flow or shallow infiltration from leaking utilities or drainage structures may have a particularly damaging effect, and the measures discussed in Chapter 3 must be applied. Depending on the character of the retained soil and the speed with which excavation is to be accomplished, predrainage to lower the surrounding water table may be necessary to prevent loss of ground. In cohesionless or nearly cohesionless soils, predrainage is essential. If this is not done it may be necessary to slow the speed of excavation so that drainage takes place concurrently.

2.   Driven soldier piles often reach a position differing from their intended location. The support system and wall design must therefore be able to accommodate this practical misalignment from design location. When cohesionless materials or soft clays are pre-excavated to better control positioning, specifications should require use of mud slurry, a casing with an auger over-cutter or some other acceptable means to prevent sloughing of the walls of the hole. It is essential that the hole be backfilled

completely with an incompressible and lightly cemented mixture after the soldier pile is installed.

3. Lagging installation must be done carefully and systematically to control loss of ground. In clay at depths where the ratio of overburden pressure to shear strength exceeds about six, plastic yield can occur at the unsupported opening. In non-plastic silt or cohesionless sand, predrainage is essential to prevent a run of material. In any friable soil, backpacking behind the boards and the placing of a porous but coherent material in the opening between the boards is essential to control ravelling and ground loss. Sand-cement grout is used often where a porous material is not required.

5.4   Cylinder Walls

Cylinder walls consist of an array of cylindrical caissons installed by a variety of methods and usually consisting of concrete or mixed-in-place soil-cement, positioned so closely as to form an essentially continuous wall. Depending on the stiffness of individual cylinders, such a wall may be rigid enough to support lateral loads with little deflection. It is seldom perfectly watertight since openings between adjacent cylinders may be large enough that retained soil can ravel through them. Then it may be necessary to grout outside the wall between cylinders or to caulk or lag the intervening space at the front face. In addition to their advantages of rigidity and their adaptability to an irregular installation arrangement, they can be utilized in a variety of ground conditions, including installation through rock-like materials by means of churn drilling or rotary coring. An alternative procedure involves hand-dug braced pits of rectangular section which may be combined in a wall to serve for both lateral restraint and underpinning. Depending on the character of the ground, the pits may form a continuous or intermittent wall. The general categories of cylinder walls discussed in the following subsections are distinguished according to the method of installation.

5.4.1 Small Diameter Piles. These are ordinarily formed using hollow-stem augers with diameters ranging from about 12 to 18 inches (0.3 to 0.45 m). Hence, they often are termed auger cast piles, and can be used for vertical as well as lateral support. The procedure is to install alternate primary piles, advancing the auger to the intended bottom position. Then after injecting a mixture of cement, sand and water through the hollow-stem, the remaining piles are placed at intermediate positions. Piles can be arranged in one or more lines as necessary to achieve the desired watertightness or stiffness. Immediately following concrete placing, a reinforcing cage or structural section is embedded in the fresh mortar.

5.4.2 Large Diameter Cylinders. Cylinders ranging from about 2 to 4 feet (0.6 to 1.2m) in diameter are frequently installed in a single line by a caisson rotary drilling machine. Depending upon ground conditions the excavation can be made with or without casing either in the dry or in a slurry-filled hole. The reinforcing cage or structural section may be placed before or after concrete placement. The caissons can be arrayed with the intention of applying gunite, shotcrete or lagging in the space between the completed units to form a wall. Alternatively, a slight overlap of piles produces a "secant" wall which is formed using a rotating casing with a cutting edge on its bottom. The cutting edge actually intersects the perimeter of previously placed piles on either side.

5.4.3 Special Conditions. The cylinder pile technique can be adapted for an irregularly aligned wall of an excavation and can be used in special cases for protection of adjacent structures without underpinning. The specification requirement for stabilization of the sides of the holes during their excavation is important, particularly where the caisson is used for protection of adjacent structures. Special posting and support for wales of the bracing system must be considered. Required treatment of the possible imperfections and leaks in the wall should be described in the specifications. It may be expedient to stipulate that grout holes be placed and grouting carried out behind the wall for a certain estimated percentage of the number of joints between cylinders.

5.5   Concrete Diaphragm Walls

A concrete diaphragm wall, or slurry wall, ordinarily is formed in a trench excavated in short sections with sides supported by viscous mud slurry. Concrete is placed in the trench bottom by tremie methods, displacing the mud slurry upward. Reinforcement of the wall is by vertical rolled steel sections or precast concrete members or by cages of reinforcing steel. Recent developments include the use of precast concrete panels lowered into the bentonite slurry. In American practice there appears to be more extensive use of steel soldier piles in the wall than other alternative methods of reinforcing. Either procedure can form the permanent foundation wall for a structure. Slurry trench construction is subject to a number of proprietary and special procedures for digging the trench and for keying adjoining segments of the wall. The wall can be made as strong and rigid as necessary by selecting its thickness and the character of the reinforcing elements which are placed within it. Since a concrete diaphragm wall can be made more rigid than either a soldier pile or interlocking steel sheetpile wall, it has been utilized in lieu of underpinning for protection of adjacent properties against movement.

Stability of the slurry-filled trench panel is a critical consideration for successful construction of slurry walls. A number of factors appear to contribute to stability of short panels in situations that are theoretically difficult to explain. This stability is related to the density of the slurry, elevation of the slurry surface compared to the outside groundwater, arching around panels of relatively short length and the effect of slurry penetration into soil on the panel walls. In highly pervious soils special admixtures may be placed in the slurry to act as plugging agents in the natural ground. In the presence of salty groundwater which would flocculate clay particles, special bentonite mixtures may be needed to avoid precipitation of the slurry.

Usually the panel width, the general character and density of the slurry, and the type and positioning of wall reinforcement are matters to be treated in specifications depending on the planned use of the slurry trench wall. Wall thicknesses are usually taken at 2, 2.5 or 3 feet (0.6, 0.75 or 0.9m) with 2 feet (0.6m) the practical minimum for excavation. Concrete placement by tremie methods requires the usual controls over the quality of construction and special care in planning the reinforcing so as to permit a satisfactory encasement by the concrete. It is important to provide for removal of coarse residue from the soil being removed which concentrates in the lower corners of the panel excavation. Details of design and construction are treated at length in References 5-1 and 5-2.

5.6   Special Wall Types

There are various methods of forming a support wall, some of them proprietary in nature, which can be used only under conditions of a royalty payment or by engaging a specialty subcontractor. Some examples are listed below, but they are by no means limited to these procedures. New methods and improvements are constantly being developed to reflect technological innovations.

1.   Freezing Procedures. An internal wall and support system may be formed by ground which has been frozen by circulating a coolant through embedded refrigeration pipes. In essence, a gravity wall of frozen ground with substantial shear strength is constructed. It may require no internal bracing. This method is particularly adaptable in certain soils where the cost of freezing is favorable. There are certain problems with ground heave that must be controlled. If ice lenses form there may be an increment of settlement on thawing that must be reckoned with. In silt or clay with water content significantly above the plastic limit, freezing can segregate water from the soil into ice lenses. Upon thawing an increment of compression can occur as the melt water is expelled. This can cause a dramatic increase in settlement in and around the thawing zone at an unexpectedly late stage of construction.

2.   Stabilized Ground. A wall may be formed by stabilization methods such as grouting. As with freezing, these procedures usually produce an increase in shear strength and a decrease in permeability. Potential for creep deformation is a concern especially where the grouted soil mass is exposed by the excavation. Testing and analysis may be needed to assess the potential for creep deformation. Grouting may be used with conventional

walls to control running soil or to seal openings or leaks. The "compaction" grouting procedure has been utilized as a means of offsetting loss of ground created by an excavation, Reference 5-3.

3. Pneumatically-Applied Linings. "Gunite" or "shotcrete", which constitute cement-aggregate mixtures of various gradations, can be sprayed as lagging between soldier piles or in connection with rock bolts and wire mesh to stabilize excavations in rock or stiff soils. They can also be used to control ravelling. In soil, this basic technique is used with closely spaced reinforcing bars. Typically, the soil must be relatively competent and the groundwater should be below the excavation bottom.

4. Reinforcement. A proprietary method has been developed involving small diameter piles, "micro-piles", installed by advancing a casing by rotary drilling after which a reinforcing bar and concrete are placed inside the casing. An array of these piles drilled at various angles in effect provides reinforcing dowels which can form a retaining barrier in relatively stable soil.

5. Soil-Cement Walls. A proprietary method is available for mixing in-situ soils with cement grouts or other additives. Vertical members, such as H-piles, are installed on a regular spacing within the soil-cement columns to provide structural reinforcement. The strength of the soil-cement depends on the type and properties of the in-situ soil, mixing ratio and injection ratio of the cement, and installation procedures such as the time and completeness of soil mixing.

6. Soil Nailing. This procedure is used for in-situ soil stabilization by installing rebar or other bar sections in small diameter holes [typically 4 to 12 inches (100 to 300mm) in diameter] drilled or augered into the excavation face. The bars are grouted and wire mesh is often attached to the bars, after which the excavation face is provided with a shotcrete lining. Alternative forms of facing and reinforcing elements have also be used. Soil nailing has been applied in a variety of soils, but typically is most amenable to stiff cohesive or cemented granular soils above the water table. The procedure has gained wide usage in recent years.

5.7   Selection of Wall Type

Selection of the wall type is influenced by factors related to the general setting and environment, type of permanent structure to be built, the appropriate or allowable methods of construction, code restrictions or local practices, and overall economy. Frequently, one specific wall and support method is not appropriate in a complex project. Accordingly, hybrid systems are planned or used eventually to conform to changes from anticipated subsurface conditions or performance requirements, or in an effort to reduce cost.

An important aspect of wall selection is the control of ground movements adjacent to the braced or tied-back excavation. Especially in urban and suburban environments, this aspect of wall performance can be critical for a successful project. Chapter 8 deals with ground movement control, and reference should be made to this chapter for a treatment of wall characteristics in relation to soil displacements.

5.8   Design Loads and Structural Requirements

5.8.1 General Design Considerations. Design loads for checking shear and flexural stresses in the support wall can differ from those used in selecting main bracing members. The main bracing members inevitably will have to accommodate the total maximum horizontal load on the wall which can be reasonably well estimated. Structural design of the wall is concerned chiefly with flexural stress produced by either positive moments between braces or negative moments at brace points. Soil arching shifts pressures away from the point of maximum deflection and reduces these moments from values computed on the basis of uniformly distributed pressures. Ordinates of the pressure diagram to be applied to braces or tiebacks are determined according to provisions of Chapter 4.

Loads used to design main internal bracing or tieback members should be selected conservatively to avoid catastrophic collapse of support systems resulting from an isolated case of overloading. Wall elements, however, may be overstressed locally and occasionally with less dramatic consequences and therefore can be less conservatively designed. Care should be exercised in the determination of negative moment in the cantilever sections above the upper support level. In this instance, a concentrated surcharge or presence of a berm above the top of the wall may create a critical condition for bending of wall elements at the uppermost support.

Vertical wall elements may be designed considering that they are continuous or hinged across horizontal points of support, or that there is some amount of fixed-end moment over the support. While ordinarily there is no doubt of the actual structural continuity across the support, a reduction of negative moment at this point may be caused by inward movement of the brace or flexure of the wall itself during excavation. The assumption as to the degree of continuity must be based on an appraisal of the probable yielding at the support level. For support systems where wales are not utilized, such as a slurry trench wall with soldier piles at the ends of short panels, the panel can be designed as a continuous two-way slab with an appropriate system of edge support.

A special problem arises where wall elements reach refusal on a rock surface which lies above the subgrade of the interior excavation. Then the rock surface must be stabilized by rock anchors, near vertical or high-angle dowels, wire mesh, pneumatically-applied cement mortar or concrete in some combination. It is particularly important that support of the rock accommodate the vertical load that is applied at the top of the rock face by the wall elements.

5.8.2 Working Stresses and Safety Factor. Structural design of wall elements should conform to the applicable specifications of the American Concrete Institute and the American Institute of Steel Construction. The choice of safety factors relates directly to the conservatism embodied in the loading diagram (see Chapter 4). It is, therefore, inappropriate to establish rigid rules relating to the allowable stress level.

In general, elements of the support system may be permitted to be loaded to higher stress levels than the ordinary allowable values provided by the reference standards for permanent construction. To evaluate positive moments in wall elements it is reasonable to permit flexural stresses in the range of approximately 1.1 to 1.3 times ordinarily allowable stresses or to adopt a reduction in design loads. However, the wall elements subject to cantilever loading above the uppermost supports should not be stressed above the normal allowable level.

The effect of temperature changes or freezing of the retained soil should be considered in the wall design. In northern climates, it can be of utmost importance to guard against the intense loading produced by build-up of ice behind the upper cantilever section of the wall. Preventive measures include insulation of the wall, monitoring of movements and temperature at critical stages, and prevention of water infiltration behind the upper part of the wall. Conversely, a drastic rise of temperature can cause distress in slender elements of the bracing system.

5.9   Wall Specification Requirements

Specifications dealing with the support wall can be of either the performance or method type. Performance specifications state only the essential function that the wall is to perform. Method specifications describe in detail the properties, characteristics, and construction methods of an acceptable wall. A method specification could include the following information for bidders concerning the wall elements of the support system:

1.   Basic loading requirements for structural design would include the pressure diagram for wall elements, minimum surcharge loads and acceptable methods of computation of stresses in wall elements.

82

2. Requirements for wall geometry may include an acceptable elevation range for its top, minimum penetration for its tip, and allowable deviation in plan position with respect to the permanent structure. The degree of watertightness or the permissible exterior drawdown should also be specified.

3. It may be desirable to stipulate maximum tolerable movement of the wall, of the retained soil or of an adjacent exterior structure. Methods that might be utilized in monitoring movements to establish conformance with these requirements should be made clear to bidders.

4. Installation of wall elements may involve limitations on construction procedures. Noise and vibration control, restrictions on pile driving methods and hours of work, assignment of working space, provisions for stockpiling materials and the like should be specified when they impose limitations, since they can have an important effect on contractors' costs.

5. Subsurface conditions which may influence wall installation should be carefully described in specifications or in information available to bidders. These would include evidence of the presence of obstructions, a rock level above the base of excavation, fill containing obstructions, rubble or old building materials at the surface, and, most particularly, the presence of utilities which need to be capped, intercepted, or controlled. It may be desirable to stipulate that a trench be excavated through surface materials which contain obstructions or utilities to identify their position before wall installation commences.

5.10 References

5-1 Xanthakos, P.P. (1979). *Slurry Walls*. McGraw-Hill Book Company, New York, NY.

5-2 Dennis, B., ed., (1980). "Slurry Walls for Underground Transportation Facilities", *Report No. FHWA-TS-80-221*, U.S. Department of Transportation, Federal Highway Administration, Washington, DC.

5-3 Warner, J. and Brown, D.R. (1974). "Planning and Performing Compaction Grouting," *Journal of the Soil Mechanics and Foundations Division*, ASCE, 100(GT6), 653-666.

CHAPTER 6: INTERNAL SUPPORT SYSTEM

6.1   Introduction

Excavation support may be provided by a system of internal braces. Bracing members may act in compression, bending, or a combined state of stress. In addition, the system may be supplemented by interior earth berms which provide additional lateral support. Sometimes the wall may act alone as a cantilever restrained only by the earth mass in which its lower portion is embedded.

Use of internal bracing has decreased in recent years, primarily because tie-backs and other forms of tension support have offered attractive alternative schemes. Nevertheless, internal braces remain an important means of resisting loads, and are advantageous in the following situations:

1.   Where the presence of poor soil conditions, buried utilities, adjacent structures, or difficulties of obtaining easements weigh against the use of tiebacks which must be drilled and anchored outside the wall.

2.   When an excavation is relatively narrow (such as a cut-and-cover subway excavation) so that cross bracing from wall to wall is practical and economical.

3.   When an excavation is small in plan dimension so that ring beams can be utilized to transfer loads directly from wall to wall.

4.   In urban areas where decking must be provided at the surface; or structures where design of the permanent floor can incorporate the temporary braces.

5.   When soil or groundwater conditions prevent the use of a simple cantilever wall.

Most of the disadvantages of internally braced systems are economic ones, primarily:

1.   Excavation is slower and more expensive between large steel bracing members.

2.   Internal bracing systems may be costly, both in terms of material (steel) and labor (welding).

84

6.2   Components of Internal Support Systems

Where there are competing alternatives, the choice among internal support methods is influenced chiefly by two factors: comparative costs and the stiffness or stability to be provided by the entire excavation support system. Bracing should be designed with construction efficiency and the final structure in mind, and to function with minimum interference with construction activities. In the following subsections the system components are discussed in relation to their typical applications. They are illustrated by photographs in Figures 6-1, 6-2, 6-3, and 6-4.

6.2.1 Horizontal Tiers of Bracing. The "cross-lot" bracing system consists of three general components which transfer lateral loads from wall to wall of the excavation:

1.   Wales are often wide-flange or H-shapes placed with their web horizontal, and are usually continuous over the wall elements. They serve as beams between struts and provide shear resistance at strut connections. Wales frequently carry combined axial and flexural stresses when compression loads are imposed from walls abutting their ends. The spacing of wales in vertical tiers is controlled by stiffness of wall members and clearances required to build the permanent structure within the excavation. Brackets on the walls and/or vertical posting between successive tiers are usually required to support the wales and to resist deadweight bending. Blocking between wales and the wall may be accomplished by a variety of means utilizing steel, concrete or wood bearing blocks, plates, and driven wedges of steel or wood.

2.   Struts are usually wide-flange, H, or pipe shapes which transfer load from wall to wall in direct compression. Bending is produced by their own weight, by eccentricity of load application at their ends, temperature changes or direct loading of construction equipment and materials. Their optimum layout is determined by the width of the retaining system, potential interference with excavation and construction activities, and economics of wale and strut sizes. Their resistance to secondary stresses and accidental damage during construction can be as important as their capacity to carry primary loads. Cross-lot struts are usually placed perpendicular to the walls, or, at reasonable lengths, installed diagonally across the corners. In some cases such diagonal struts may serve internal bracing requirements entirely.

When ground movement control is important, preloading struts or braces is an effective way to reduce lateral wall displacement. Preloading often is performed by hydraulic jacking, with metal shims or plates inserted between the wale and braces to lock in the load. Braces can be installed in tight contact with the excavation walls and preloaded at low to moderate

Figure 6-1
Soldier Pile Wall Supported by Raking Bracing

86

Figure 6-2
Soldier Pile Wall for Subway Supported by "Cross-lot" Struts

87

Figure 6-3
Combination of Diagonal Corner Braces and Raking Braces
(viewed from outside excavation)

Figure 6-4
Combination of Diagonal Corner Braces and Raking Braces
(viewed from inside excavation)

levels with driven pairs of steel wedges. Alternatively, when some ground movement is acceptable, braces can be cut to fit and welded or spliced to the wales as conditions dictate.

3. Secondary Members include angle lacing and tie beams to ensure elastic stability and to reduce slenderness ratios both in horizontal and vertical planes, posting to provide vertical support, and most importantly, the connections between wales and struts.

6.2.2 Inclined Braces. Rakers commonly are employed where cross-lot bracing is uneconomical or not feasible because of excavation width or construction interference. Compared to cross-lot bracing, additional opportunities for movement are caused by the difficulty in obtaining a rigid reaction at the heel of the inclined member. Problems are posed by the upward component of thrust at the wale and the fact that interior excavation leaves only a berm of triangular or trapezoidal cross-section against the wall prior to the time the bracing is placed. This system is illustrated in Figure 6-1 and includes:

1. Raker. The raker is the inclined member slanting downward from the wale to an interior reaction. Steel wide-flange, H or pipe sections or, occasionally, square wood members are utilized. For equivalent lateral load in the same size excavation a greater cross-sectional area must be provided for the raker than with cross-lot bracing because the lateral earth pressure is resisted only by the horizontal component of the force in the inclined member. The axial load depends on the inclination of the raker and is usually 10 to 20 percent greater than the horizontal load.

2. Reaction. The reaction to the inclined brace load ordinarily is provided by one of the following elements: a temporary pad or inclined footing of concrete or timbers; permanent cast footings or mat foundations; heel blocks or stub beams embedded in mat sections; pile caps or batter pile frames. Reaction elements may be placed as part of the permanent construction inside the berms. The reaction must resist both vertical and horizontal loads with appropriate limitation of movement. Interior subgrade conditions may restrict the means of providing a reaction and often a judgement must be made of its stiffness without a quantitative evaluation of the movement expected. Use of permanent foundation elements may dictate a special sequence of excavation and concrete placement as well as restrict their use. Compression members can be installed between symmetrically positioned reaction pads to transmit horizontal loads from wall to wall of the excavation or between several lines of individual footings.

3. Special Requirements. The raker reacts against the wale with an upward-directed force component so that the wale must be bracketed to accommodate a twist and vertical-directed shear. These forces are in

addition to wale loads present in the cross-lot system. The bracket is usually provided by a structural tee section. To reduce l/r values, a raker may have to be posted by piles driven through the beam plus lateral tie beams, or by some other means. Preservation of lateral resistance in the berm during placing of the braces often requires a tightly controlled sequence of excavation wherein rakers may even be placed in trenches or slots cut in the sloping face of a berm. These activities, however, inevitably weaken the berm.

6.2.3 Special Methods of Support. These comprise features that may be incorporated in systems similar to the two main methods discussed above and include the following:

1. Permanent Floor Systems of the below-ground structure are used as horizontal braces for excavation walls in special circumstances. Similarly, steel cross-lot braces may be incorporated later in the floor system. A special example is the installation of floor members against previously-constructed walls as excavation proceeds downward for a deep basement. Excavation support combined with permanent floors can provide important economy but may also greatly complicate construction and usually requires accurate control of loads in the braces. Furthermore, it reduces the contractor's options to utilize systems of his own choosing.

2. Top-Down Construction makes use of concrete diaphragm walls, load bearing elements, and permanent floors for the excavation support system. Initially, cast-in-place concrete diaphragm walls are installed around the perimeter of the excavation to act as permanent foundation walls. Load bearing elements in the form of drilled caissons or slurry wall segments, are installed to support interior columns. The ground floor slab and subsequent lower floors are then cast in sequence from the top down as soil below each slab is excavated in a mining operation. The excavation and support sequence can be performed simultaneously with construction of the superstructure, and this application is referred to as up-down construction. The approach may save substantial construction time which offsets additional costs associated with the special excavation and support procedures. Because foundation elements for column support must be installed before excavation, certain types of foundations cannot be used with maximum economy, including spread footings, mats, and pile clusters.

3. Interior Berms can be employed as the sole means of interior support of cantilever walls, usually where excavation depth is not great and there is ample peripheral space between the wall and structures to be placed in the interior. A berm can be economically attractive, but results in a yielding support. Berms create a problem if limitation of movement or stability is critical since an accurate appraisal of their performance requires a

moderately sophisticated investigation and analysis. References 6-1 and 6-2 illustrate some difficulties with berm resistance in excavation support.

4. Ring Beams or Wale Support are most frequently employed in shafts of circular, square, or near-square section without struts. In circular shafts the ring beams almost invariably are designed to work in compression only, although in some cases consideration of combined stresses is important. In square-cornered excavations the wales will be subjected to significant combined stresses and this generally limits the length of the wall face over which they can be employed. One means to avoid the need for cross-lot struts or rakers is to construct an upper wale of great stiffness with reinforced concrete, installing corner braces designed to permit fairly long dimensions between brace points, thus minimizing interference with interior construction.

5. Timber Material was used rather extensively in the past for internal bracing systems and for the wall itself since it is relatively lightweight and economical and is readily handled and shaped. It is, however, much less rigid than steel and is now chiefly limited to narrow trenches or to shafts and pits in situations where the support system is not selected by engineering design. A detailed study of lumber used for trench bracing is given in Reference 6-3. Timber rakers are occasionally used with steel wales.

6. Special Prefabricated Systems comprise relatively light steel members, often circular or box shapes which are rigid for use in trench bracing. Such systems involve a selection by empirical rules or the experience of the contractor and usually are not engineered for specific applications.

6.3   Loads in the Support System

Procedures for estimating lateral pressures applied to the excavation support system are given in Chapter 4. Application of the loading diagrams to wall units is discussed in Chapter 5. This section is concerned with use of the pressure diagrams thus derived for the design or selection of wales, main bracing members and secondary members of the internal bracing system.

Apart from factors enumerated in Chapter 4 that influence lateral pressures on the wall, design loads in individual tiers of bracing are controlled by: the method and sequence of construction, stiffness of the retaining system as a whole, and certain fundamental simplifying assumptions of the performance of the support system. The following factors should be considered in assessing design loads applied to the support system:

1. Distribution of Lateral Loads to the Wales. To compute line loads acting along the wale at any tier of braces, the wall is usually assumed as simply supported at the brace level, except that the wall is taken as a cantilever above the upper brace. As discussed below, a degree of continuity over the lowest brace may be assumed for some conditions.

2. Surcharge. It is of vital importance that a realistic design surcharge be applied to the wall. The surcharge must account for possible concentrations of construction materials and heavy equipment servicing the excavation activity, the effects of earth slopes cut above the top of wall, or the pressures applied by existing adjacent structures.

3. Maximum Brace Load. Maximum live load in the wale and strut can occur at an intermediate construction stage just before placing the brace next lowest to the tier being considered. This condition should be checked assuming an equivalent point of simple support at or below the base of the excavation with lateral pressures transferred by that span to the brace above.

4. Effectiveness of an Interior Berm. In evaluating brace loads during construction it is important to appraise realistically the resistance provided by an interior berm. Theoretical passive resistance may be computed as in Chapter 4, but a liberal safety factor should be applied to these theoretical values to ensure that the desired reacting force will be mobilized by movements which are tolerable. This safety factor ordinarily ranges from 1.25 to 1.75. The specific choice depends on factors such as: the potential for creep of the berm material; the threat of sloughing, erosion or over-steepening of the berm's inner face; time of exposure; and effects of construction operations, particularly interior pile driving, trenching on the berm slope, and installation of drilled piers.

5. Overall Stability Conditions. As explained in Chapter 4, safety must be provided against undermining the wall and heaving of the excavation base by overall mass movement. In the most common case where excavation encounters progressively more compact soil as depth increases and where groundwater is properly controlled, the base is stable against overall shear failure. In this case, wall units need to extend below the base only to a sufficient depth to provide vertical support for the wall and to obtain any necessary reaction for the bottom segment of the wall below the lowest brace. Loads on the lowest brace can then be taken as those produced by net pressures on the lowest part of the wall acting as a cantilever bending about the lowest brace. If penetration of the wall below the interior subgrade is a limited amount, this cantilever can be partially supported by passive resistance of the interior soil below the subgrade. If penetration of the wall is substantial it may be judged that a point of equivalent simple

93

support has been formed in the wall at or just below the base of excavation by mobilizing the passive resistance on the more deeply embedded section.

6. Overall Unstable Conditions. If there is a threat of failure or mass movement beneath the base of the excavation, the wall units may have to be extended to greater depths to reach a compact stratum and engage its higher lateral resistance. Loads on the lowest tier of bracing could be increased substantially over the ordinary design pressures by arching upward of pressures from the unstable zone. The magnitude of these additional forces should be evaluated as outlined in Chapter 4.

7. Seepage Effects. To avoid excessive conservatism in assessing ground water pressures, it is useful to analyze for water pressures which accompany seepage flow into the excavation beneath the bottom of the wall. Ordinarily at the buried tip of the wall the water pressures on both sides will be equal and no net unbalanced water pressure will be acting across the wall at that level. The consequent reduction in net water pressure acting upon the buried part of the wall and the portion of wall below the lowest brace should be taken into account in evaluating the lowest brace load and negative moments in the wall units at or below the lowest brace.

8. Interior Subgrade Conditions. Conditions of the interior excavated surface at an intermediate stage or at the final subgrade can affect the interior soil's passive resistance. Drainage ditches cut near the toe of berms, over-excavation for pits, driving of bearing piles, installation of drilled piers or uncontrolled upward seepage may reduce passive resistance. These aspects of the construction can increase net forces on the lowest tier of bracing.

9. Rebracing Conditions. After the basic support system and the excavation are complete, rebracing is sometimes necessary to reposition members which interfere with pile driving or subsequent permanent construction. In deep excavations, the circumstances surrounding rebracing may create maximum brace loads if large clear heights are desired for the foundation wall pours. Therefore, a probable sequence of rebracing should be taken into consideration in evaluating brace loads. The mat or lowest permanent foundation construction can provide nearly fixed-end moment resistance to bending of the wall at the level where it is poured against the wall. The degree of fixity which this provides may be taken into account in analyzing loads carried to the next tier of bracing left in place above.

6.4 Structural Design Criteria

There is no consensus on detailed criteria for bracing system design. Numerous judgmental factors are involved in selecting the lateral pressure diagram, assessing

94

the passive resistance and choosing the maximum brace load. There is also an important question as to stress levels that may be employed in various units of the support system. For instance, in steel bracing systems at one extreme all stresses in all elements would simply be kept at allowable working stresses defined by the American Institute of Steel Construction (AISC) specifications for permanent construction. At the other extreme, the stress level for maximum load under temporary conditions would be permitted to reach values roughly midway between AISC allowable values and the yield point.

In the future, more use may be made of "load factors" applied to the various component portions of the lateral total pressures acting on the support system. This would be associated with an analysis for ultimate stresses in the excavation support elements. The load factor procedure has the disadvantage of obscuring the conventional and familiar values of the lateral total pressure. Nevertheless, account should be taken in design of the predictability or uncertainties of the loading, the degree of risk and the consequence of distress in assigning allowable stresses in the retaining system.

Compression members in almost all internal bracing systems (even ring beams under symmetrical loading) are subject to combined stresses. The designer must check the wales and struts for this condition. The connections, particularly between wales and both struts and rakers, are subject to combinations of forces for which the precise effect is beyond analysis. Actual stress levels are a function of construction details and workmanship which cannot be predicted with great accuracy. Because they are critical elements in the support system, main bracing compression members should be designed more conservatively than the wall itself. The connections should probably be operating at the lowest allowable stress levels of any part of the system. In the following subsections, design aspects of the various elements of the support system are considered in more detail.

1. Design of Wales is generally controlled by bending combined with axial compression in the spans adjacent to the corners. It may be economical to decrease spacing between braces in this area to equalize stresses throughout the wale length. Continuity of the wale across brace points is normally assumed, provided the necessary flexural capacity is incorporated into wale details at splices and at the corners. It is not necessary to be conservative in evaluating the maximum positive moment between braces since the brace-wale connection unquestionably tends to provide a hard point upon which wall loads concentrate. Flexure due to weight of the wale in the vertical plane usually is ignored if braces or posts are provided at reasonable spacing or the wale is attached to the wall. With a soldier pile wall it is more economical when struts join the wale at a soldier pile because wale bending moments are thus decreased; but it is not necessary, provided torsional stresses in the wale are accounted for. With rakers, the wale connection must be placed at a soldier pile so that the upward raker

component in the wale can be resisted efficiently. Combined stresses under longitudinal compression forces applied from end walls where the wales serve as struts usually are permitted to reach between 1.0 and about 1.35 times AISC normal allowable combined stress depending on the degree of conservatism of the loads selected for design. Because combined bending-compression is critical to wale design, an important consideration is the dissipation of the axial load in the wale into the wall and thus into the retained soil. The wale-wall connections must be designed to accommodate some reasonable pattern of this dissipation.

2. Main Bracing Members. With the maximum compression load determined as described in Section 6.3, the direct stress usually is permitted to reach between 1.0 and 1.25 times AISC normal allowable direct stress considering the specific l/r provided by posting piles, tie beams or lacing. If the maximum load is computed for the condition of an intermediate stage of excavation, the stresses usually are taken toward the upper boundary of this range. Tie beams and lacing members should accommodate roughly two to four percent of the design axial load in the main member. Posts must handle the dead load of the system plus anticipated construction surcharge imposed on the braces.

3. Connections. Connections should be designed at AISC normal allowable stress to accommodate about 1.25 times the computed maximum stress in members produced by direct loads, moment or shear. Connections include stiffeners in the wales, angles or brackets which are intended to prevent twist or buckling and either provide the necessary vertical reaction to a raking brace or accommodate the dead load of the wale itself. Compression member connections in steel should be designed for shear or temporary tension equal to 10 percent of the compressive load unless the actual shear or temporary tension exceeds this amount. It is essential that the design should provide for maintaining tight contact between support members. Ordinarily this requires well-distributed blocking points placed between wale and wall or, alternatively, fitted and welded connections. During placement of nearby braces in a particular tier, wedges should be checked and re-driven and wall blockings kept snug. Particular attention should be given to the snugness of wedges which have been installed during warmer periods of the day. At a time after movements in that tier have stabilized the wale-brace connection can be tack welded.

4. Selection of Permissible Stress is a matter of judgement not subject to agreement among engineers with different viewpoints. Factors considered are the duration of least favorable loading conditions, the importance of consequences of over-stress, the potential for rapid and uncontrolled progressive failure, and, most importantly, the degree of confidence in the construction workmanship and inspection. Certainly, it is rational to load

the temporary support system at higher stress levels than would be permitted for a permanent structure. However, the system suffers from all the imperfections of on-site temporary construction, and this does not warrant disregard of normal recommended working stress limits. Conservative judgements should be focused particularly on connection details and main bracing members.

6.5   Pre-loading of Bracing Systems

In important or large-scale internally braced excavations, pre-loading of braces has become increasingly common. The purposes are to restrict movement of the wall by removing slack and some elastic deformation in the system and to provide both uniform distribution of loads in a tier of braces and a systematic distribution between tiers compatible with design assumptions. When struts are pre-loaded, care must be taken so that the wall is not over-stressed locally due to a lack of stiffness or continuity in the support system or because of the yielding nature of the retained soil.

6.5.1 Methods of Application. Some pre-load may be applied in the simplest case for cross-lot braces by driving two or three pairs of steel wedges between a pair of bearing plates where the strut and wale meet, or for rakers, at the lower end. The amount of load that can be applied in this manner generally is limited to stress increases of several kips per square inch, although somewhat higher levels can be achieved by using more expensive milled wedges. In more important cases the preload is applied by a pair of hydraulic jacks acting against brackets on individual struts or rakers. After stressing to the desired load level, the load is transferred by inserting or driving steel wedges and shims as described above. Where desirable, the brace load can be determined fairly reliably as construction proceeds with various types of strain gauges installed on the struts or rakers.

6.5.2 Selection and Maintenance of the Pre-load. Struts are typically pre-loaded to between 1/3 and 3/4 of maximum design load. The lower value is chosen when a moderate amount of movement is expected and tolerable. The higher value is selected when movement must be minimized and the support system, including the wall, is stiff and provides the necessary continuity to accept high brace loads applied in succession. High pre-loads cannot be utilized effectively if the support system is flexible; for example, if struts or rakers are placed directly against soldier piles without a wale connecting adjacent soldier piles, or where only one wall is stiff in a cross-lot system. Depending on the conditions at the time of pre-loading, some difference in value between design load and pre-load must be reserved for expected increases in temperature of the brace. Where struts will be incorporated as permanent floor beams, the selection of pre-load must leave adequate reserve capacity to accommodate the final flexural stresses in the floor. In this case, while the pre-load may be a fairly high percentage of the design lateral load, it can be working the beams at a fairly low stress level. Casting of a concrete floor over the steel braces can result in a significant temporary increase in stresses due to temperature rise as the concrete cures. Shrinkage of concrete in the final stages of curing reduces these stresses. After pre-loading, it is essential that observations be continued to check the loosening of contacts, or the distortion of connections from excessive loading, or eccentrically applied loads. The tightness of wedges should be checked by periodic driving, particularly in times when temperature is falling. Additional measures, such as painting braces with reflective colors may be used to reduce thermal effects.

6.6   Specifications and Construction Requirements.

As discussed in Chapter 1, an excavation support system is designed in most cases by a contractor selected by the owner or by an agent of the owner, who often is the structural engineer for the project. It should be recognized that the contractor-designed and consultant-designed approaches will have different implications with respect to specifications and construction requirements. In this section specifications are treated in a general way, with the recognition that design and specification responsibility must be judged on the merits of each particular case.

Specifications may be categorized as the "performance" or "method" type. In its most elaborate form, the method specification includes a completely designed and detailed excavation support system with specifications as elaborate as those for permanent construction. At the other extreme, a minimum performance specification would include merely a statement of the functions that the excavation support system must fulfill, such as, permitting the excavation to be done safely in the dry while limiting ground movements and protecting adjacent structures.

One fundamental stipulation which can inform bidders of the structural capacity required of the excavation support system is to present a basic lateral loading diagram, including both earth and water pressures for the condition of completion of the excavation. Some of the requirements suggested in Sections 6.3, 6.4 and 6.5 which describe the means by which the excavation supports will resist this basic load may be included.

As a general rule, framing specifications to achieve an apparent initial economy in the support system by omitting or de-emphasizing technical requirements may be ill-advised and finally uneconomical. Inclusion in contract documents of a complete support system design can promote a more equitable competitive bidding. Bidders do not have to include the range of contingencies necessary for their tentative design and the specified system can be used as a point of departure for their own alternatives. On the other hand, qualified specialist contractors may offer unique support systems which accomplish a specified result and effect significant savings for an owner. A method specification will limit this sort of contractor initiative. The choice between "method" or "performance" specifications is thus often best made with knowledge of the prospective bidders and their capabilities.

The specialized nature of the work for excavation support systems has contributed to the increased use of designs by qualified specialty contractors. Such an approach does not disqualify the owner's design consultant from playing a major role in the choice of systems when the application is critical in nature. As a general rule, performance specifications should be provided which detail the minimum requirements for contractor qualifications and experience. The more critical the application, the more restrictive these requirements should be.

Performance specifications for this work generally fit into one of two categories:

1. Fundamental specifications which merely state which requirements the shoring system must fulfill; such as permitting the excavation to be done safely in the dry while limiting ground movements and protecting adjacent structures. This category is generally used in less critical applications.

2. Specifications for critical applications which may provide minimum lateral loading diagrams, including both earth and water pressures for the final excavated condition. They may even address temporary conditions during the excavation phase. In particularly critical applications, some of the requirements suggested in Sections 6.3, 6.4 and 6.5 may be included which describe minimum factors of safety and the means by which the shoring will resist basic loads.

Once the performance specifications are established, a qualification procedure should be drafted. In drafting the qualification procedure, it would be advantageous to emphasize contractors with proven expertise and experience.

Regardless of the form of specifications, it is beneficial to make a series of decisions about the effects of the temporary construction on the permanent structure and to impart these decisions to the contractor. These should include judgements made on the following points:

a. The need and requirements for boxing out, cutting or removing braces that pass through permanent concrete or for casting into the concrete.

b. Requirements for removal or cutting off below final grade of wall elements, such as soldier piles, sheeting or lagging.

c. Treatment of the crucial area between the interior permanent structure and the wall, including control of backfilling, attachment and protection of waterproofing materials, and removal of the temporary structures.

d. Restrictions on the loading applied to the permanent structure as part of the support system at intermediate stages of removal of the temporary structure.

6.7 References.

6-1 Clough, G. W., and Davison, R. R. (1977). "Effects of Construction on Geotechnical Performance," *Proceedings*, 9th International Conference on Soil Mechanics and Foundation Engineering, Tokyo, Japan, Vol. 3, 15-33.

6-2 O'Rourke, T.D. (1981). "Ground Movements Caused by Braced Excavations," *Journal of Geotechnical Engineering Division*, ASCE, 107(GT9), 1159-1178.

6-3 Knab, L.I., Yokel, F.Y., Galligan, W.L., Bendtsen, B.A., and Snaft, J.F. (1980). "A Study of Lumber Used for Bracing Trenches in the United States," *NBS Building Science, Series 122*, U.S. Department of Commerce, National Bureau of Standards, Washington, DC.

CHAPTER 7: TIEBACK SUPPORT SYSTEM

7.1   Introduction

Tieback anchors are commonly used for temporary wall support on major excavation projects. This chapter discusses advantages and disadvantages of anchoring methods as well as pertinent aspects of successful installations. Many installations are designed and installed by specialty contractors using proprietary methods. Some installations, mainly permanent ones, are designed by the owner's engineer and bid competitively. In either case, it is desirable for the owner's engineer to provide oversight regarding design earth pressures and excavation effects on adjacent property, to participate in formulating performance criteria, and to establish reliable methods of evaluating tieback wall performance.

The first part of this chapter addresses tieback design. The second part addresses tieback construction. Views of typical anchored wall projects are shown in Figures 7-1 and 7-2. General summaries of practice are given in References 7-1 and 7-2. Case history information may be found in Reference 7-3.

A well-designed tieback job should be technically feasible, economical, and safe. A few fundamental questions should be addressed before deciding to use tieback anchors. Anchor capacity is developed from the resistance of surrounding ground. Most soil and rock strata, with the possible exception of loose fill, organic soils, soft clays or silty clays, are suitable for providing resistance. Nearby buildings and foundations, utilities and subways can all interfere with installation. The impact of an anchored wall on the project must be assessed, including comparisons with interior bracing schemes (which themselves may interfere with the construction operations) which likewise have advantages and disadvantages.

The designer of the tieback system should be familiar with the types of tiebacks most commonly used in the project area, and evaluate the type of tieback most favorable for site soil conditions. Many of the tieback types in current use were developed by specialty contractors for particular equipment or for specific soils. Engineers interested in specifying a tieback system should be acquainted with the contractors qualified to perform the work in the project area.

To obtain the most economical price, specialty contractors should be asked to make an anchor selection to suit the project. Thus a performance specification is most often appropriate. Because of proprietary methods and varying experience, bidders may propose different designs and may be able to offer economy not available if a completely detailed method specification is imposed.

Figure 7-1
Hybrid Excavation Support System:
Soldier Pile Wall Supported by Tiebacks (Upper Wale) and Rakers (Lower Wale)

Figure 7-2
Examples of Deep Excavation with Soldier Pile Wall and Tiebacks

7.2   Anchor Types and Selection

As illustrated in Figure 7-3, anchors for tiebacks are made in two basic configurations. The first is a cylindrical shaft anchor which mobilizes the shearing resistance of the ground around its perimeter. Such anchors are formed in holes drilled into soil or rock. Secondary or multiple stage high pressure grouting may be used to increase anchor capacity. The second basic configuration involves a bottom section which is belled to enlarge the anchor in cohesive soil or soft rock. In this case, some portion of the total capacity results from end bearing of the enlarged portion.

7.3   Design, Analysis, and Review

Tiebacks consist of a steel tendon connecting a wall to an anchor grouted in the ground. The tendon is inserted into a pre-drilled hole which may be lined with temporary steel casing. The anchor is the grouted length of a tieback which bonds to soil or rock and transmits the design load. It is important that no part of this anchor be developed in the zone between the wall and the anticipated soil failure surface. Accordingly, the upper portion of the steel tendon is not bonded to the grout. The extent of this un-bonded length is influenced by soil properties and also by mass stability analysis. The tieback or anchor angle is commonly taken as the angle between the axis of the tieback and horizontal.

7.3.1   Mass Stability Analysis. The tieback and wall, together with the retained soil, is assumed to fail along a linear, curved, or composite surface as shown in Figure 7-4. In design, a number of such trial surfaces are examined and the mass stability along the most critical surface is determined. A suitable safety factor along this surface is critical for an acceptable design. The analysis typically involves a series of trial free-bodies bounded by straight-line or curved segments. The failure geometry is chosen variously to pass through or outside the anchor. Reference 7-4 gives examples of stability analyses adapted to various tieback arrangements.

7.3.2   Tieback Layout and Arrangement. The design drawings should show the required anchor capacity, the type of anchor to be used, the type of tendon, and the manner in which the tieback will be secured to the wall. The design drawings should also show the locations of the tiebacks and the range of anchor angles. It is common practice to specify a minimum unbonded steel length corresponding to the distance inside a potential failure surface. The tiebacks should extend a sufficient length beyond this surface so that the soil mass they support forms a stable free body.

The failure surface frequently is evaluated as a Rankine wedge with modifications, as appropriate, to account for failure surfaces extending below the base of excavation, for example, see Reference 7-5. The surface often is drawn as a plane

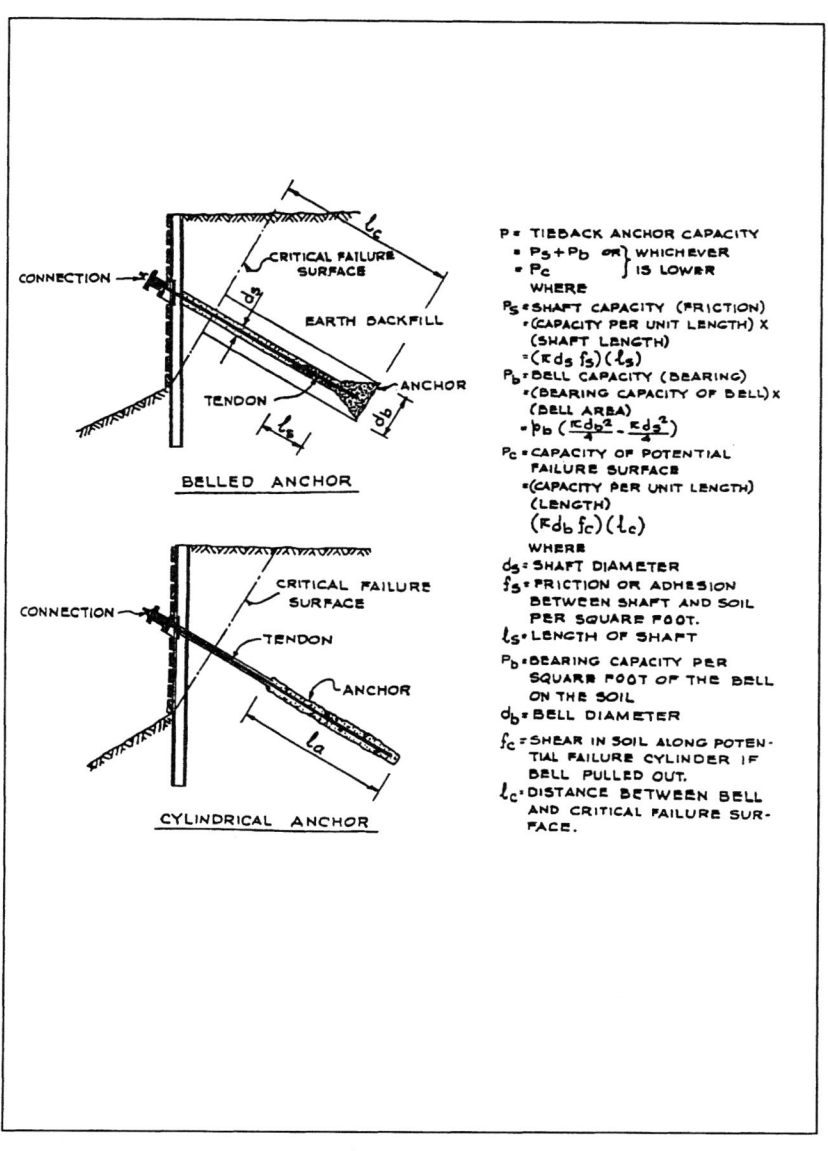

P = TIEBACK ANCHOR CAPACITY
$= P_s + P_b$ or $\left.\right\}$ WHICHEVER
$= P_c$ $\left.\right\}$ IS LOWER
WHERE
P_s = SHAFT CAPACITY (FRICTION)
= (CAPACITY PER UNIT LENGTH) X (SHAFT LENGTH)
$= (\pi d_s \, f_s)(l_s)$
P_b = BELL CAPACITY (BEARING)
= (BEARING CAPACITY OF BELL) X (BELL AREA)
$= P_b \left(\dfrac{\pi d_b^2}{4} - \dfrac{\pi d_s^2}{4}\right)$
P_c = CAPACITY OF POTENTIAL FAILURE SURFACE
= (CAPACITY PER UNIT LENGTH) (LENGTH)
$(\pi d_b \, f_c)(l_c)$
WHERE
d_s = SHAFT DIAMETER
f_s = FRICTION OR ADHESION BETWEEN SHAFT AND SOIL PER SQUARE FOOT.
l_s = LENGTH OF SHAFT
P_b = BEARING CAPACITY PER SQUARE FOOT OF THE BELL ON THE SOIL
d_b = BELL DIAMETER
f_c = SHEAR IN SOIL ALONG POTENTIAL FAILURE CYLINDER IF BELL PULLED OUT.
l_c = DISTANCE BETWEEN BELL AND CRITICAL FAILURE SURFACE.

BELLED ANCHOR

CYLINDRICAL ANCHOR

Figure 7-3
Elements of Tieback System

105

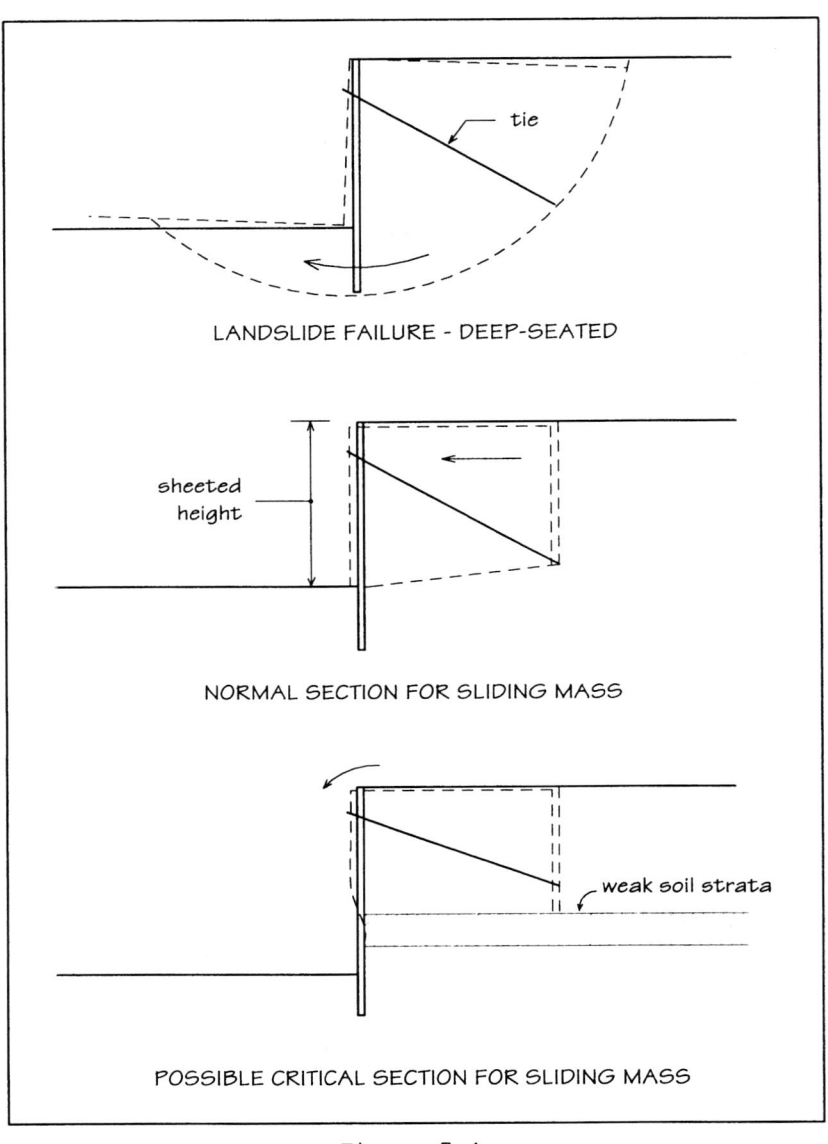

LANDSLIDE FAILURE - DEEP-SEATED

NORMAL SECTION FOR SLIDING MASS

POSSIBLE CRITICAL SECTION FOR SLIDING MASS

Figure 7-4
Failure Surfaces for Mass Stability Analysis

from the base of the excavation or some depth below the base at an angle with respect to the vertical of 30 to 40°. Consideration should be given to the soil strength properties and depth of anticipated wall movement before a failure surface is established for a specific wall geometry.

The earth pressure load diagram, available anchor capacity, and design of the support wall need to be considered in determining the optimum location of tiebacks. The fundamental design consideration is that the wall, with lateral support in selected locations, must restrain the calculated earth loads.

One often overlooked factor which affects the positioning of tiebacks is the influence of new construction just inside the support wall. The elevation of permanent construction joints, floor slabs, or the location of columns may affect the location of tiebacks which might interfere with interior wall construction. Serious failure of a retaining structure could result from the premature release of an anchor due to interference with new construction. The plans should consider such potential conflicts, and provide a procedure for prevention or resolution.

7.3.3 Capacity of Individual Tiebacks. Capacity of individual tiebacks generally is limited by either the strength of the steel tendon or the ability of the anchor to resist pullout. An optimum design would attempt to make these two capacities equal.

The tendons used in tieback anchor construction generally consist of the same types of steel used in post-tensioning applications. Tendons are designed using an allowable stress approach. Tendon design capacity is computed as:

$$t_a = AR\sigma_u \qquad (7\text{-}1)$$

in which:

t_a = tendon design capacity

A = the cross sectional area of the steel

R = the percentage of allowable stress, normally computed at 60% of the guaranteed ultimate tensile strength (GUTS) for temporary work and 50% of GUTS for permanent work

σ_u = the GUTS of the steel (usually 150 ksi for bar and 270 ksi for strand)

Anchor length is an important consideration in tieback design. Figure 7-3 shows two ways in which anchor capacity is normally calculated. A tieback typically employs a cylindrically shaped anchor the capacity of which is computed as the

107

product of anchor length and capacity per unit length. The capacity per unit length can be estimated (often by the specialty contractor) on the basis of soil boring information and experience in site conditions. It may also be back-calculated from anchor pullout tests performed on site. A preliminary check of capacity is represented by the following formula:

$$P_u = \pi \, d \, l \, f \tag{7-2}$$

where:

P_u = the ultimate anchor capacity

d = nominal diameter of the grouted anchor length

l = the length of the anchor

f = the unit pullout resistance at the interface between the anchor and the soil or rock

The resistance to pullout may be increased in soil by use of modified grouting methods. However, resistance may be limited by the shearing resistance of the adjacent soil.

7.3.4 Estimating Pullout Resistance. Methods of estimating unit resistance are available in the literature (References 7-2 and 7-4). Methods are typically based on those used for estimating available skin friction on piles. For cohesive soils, a constant value equal to a portion of the peak shear strength is used. For cohesionless soils, pullout resistance depends on soil density, friction angle and effective overburden pressure, expressed in terms of average depth of the anchor below ground surface, with some allowance for water table conditions. However, because disturbance to the soil is greatly influenced by drilling and installation methods, estimates of pullout resistance are only approximate. Where grout is injected under pressure to form the anchor, field measurements show that the capacity per unit length can be substantially higher than calculations based on ordinary skin friction and vertical effective stresses would permit. Measurements of capacity per unit length for pressure injected anchors in various granular soils show values in the range of 75 to 300 kN/m (Reference 7-6).

An important advantage with tiebacks is the opportunity to perform field pullout tests to verify estimated load capacity. This testing guarantees that a certain minimum factor of safety is available before excavation continues. It is typically recommended that every anchor be tested; temporary anchors to a minimum 133% of design load and permanent anchors to 150% of design load. In cohesive soils, additional testing should be performed to satisfy allowable creep criteria, especially for permanent installations.

7.3.5 <u>Vertical Component of Tieback Force</u>. Non-horizontal tiebacks exert a downward force component on the wales and wall. For this reason, tiebacks with modest inclinations are usually preferable to steep ones. To develop a given horizontal force, more tiebacks installed at steep angles are required than if tiebacks are installed at shallow angles. Likewise the vertical component of a steeper tieback is greater. When the tied-back wall settles, less horizontal movement occurs with the flatter tiebacks. Vertical effects are minimized if the sheeting wall is adequate to transmit the vertical loads to soil beneath the excavation, or if shear between the back of the sheeting and retained soil is adequate to provide the required vertical reaction.

If tiebacks are anchored into rock or firm soils underlying soft clay, base stability of the excavation may be affected by the vertical anchor force components. Procedures for evaluating the reduction in safety factor against base failure have been proposed for this situation (Reference 7-9), and compared with the field performance of several excavations in clay with tieback anchors and underlying firm ground.

7.3.6 <u>Permanent versus Temporary Tieback Installations</u>. Tiebacks used for temporary support have a fairly short life span -- usually from a few months to several years. No special corrosion protection is usually required (in the unbonded length or the anchorage length) for these tiebacks unless in a highly corrosive environment. For the unbonded length, the grease and plastic sheathing used to debond the tendon from the grout inhibit steel corrosion. The anchorage length is protected by full grout encasement. The connection at the wall and the upper several feet of a temporary tieback require attention. Standard practice dictates that the wall connection be protected for the first few feet behind the wall, particularly in the case of bar tendons which may corrode rapidly in zones where bending stresses are concentrated.

When installing permanent tiebacks, it is necessary to provide a more substantial level of corrosion protection. The upper anchorage connection should be adequately encased in concrete or covered with a grease (or grout) filled metal cap. The unbonded length should be coated with special grease chosen for its permanent and corrosion-inhibiting properties. The upper portion of the unbonded length which may be exposed to oxygen and water should be encased in a metal or plastic sleeve. In highly aggressive conditions (such as low pH, high sulfide/ chloride content or stray ground currents), a more sophisticated protection system should be employed. One such system is complete encapsulation of the tendon in a corrugated plastic sheath. The corrugations provide mechanical stress transfer between the grout inside and the grout outside of the sheath. The plastic itself isolates the steel from the aggressive conditions present in the soil and ground water. Various configurations and protective measures for permanent tiebacks are described in References 7-8 through 7-10.

7.4   Tieback Anchor Installation

Anchor hole drilling should be performed using a method which permits reasonably accurate location control and provides the required holding capacity. Many drilling techniques are available which satisfy these criteria. Nevertheless, care must be exercised in evaluating the drilling procedures with respect to ground disturbance. Examples of methods with the potential for excessive ground disturbance are:

1.   Auger equipment used to drill at a shallow angle in cohesionless soil.

2.   Vibratory or heavy impact hammers which are used to drive temporary steel casing in loose sand.

3.   Rotary percussion drilling methods in sand where air is introduced through a drill bit in front of the casing to flush soil to the top of the hole.

To promote efficient use of drilling or grouting equipment, it generally is advantageous to excavate only a relatively small distance below a level of tiebacks before installation. Accordingly, there is an incentive to install tiebacks in a timely fashion and therefore limit the depth of excavation below the lowest level of support. This promotes lateral stability and helps to control wall movements.

Every tieback is usually proof-tested by pre-tensioning the tieback beyond the design loading. Typical testing procedures and equipment are described in the following sections.

7.4.1 Proof Testing. Proof testing is performed by staged application of load to the tieback with a hydraulic jack until reaching test load. The load is then reduced to some lock-off load, generally 75% to 100% of design load. A typical test set-up is illustrated by the photograph in Figure 7-5. A normal loading series might involve four to six equal increments at about 25% of design load. Loads and displacements associated with a typical proof test are shown in Figure 7-6.

The proof test is a rapid and routine procedure, generally of 15 to 30 minutes in duration. To minimize wall deflections, it is advantageous for tendons to be prestressed rather than allowed to elongate with increasing load as excavation proceeds. The proof test is useful in that it not only verifies the capacity of a tieback, but also pre-stresses the tendon and support system.

7.4.2 Jacking and Measurement Procedures. Hollow-center hydraulic jacks bear against the wall and are used to apply loads to the tiebacks. Jacks should have a

Figure 7-5
Typical Setup for Tensioning Tieback Consisting of High-Strength Threaded Bar

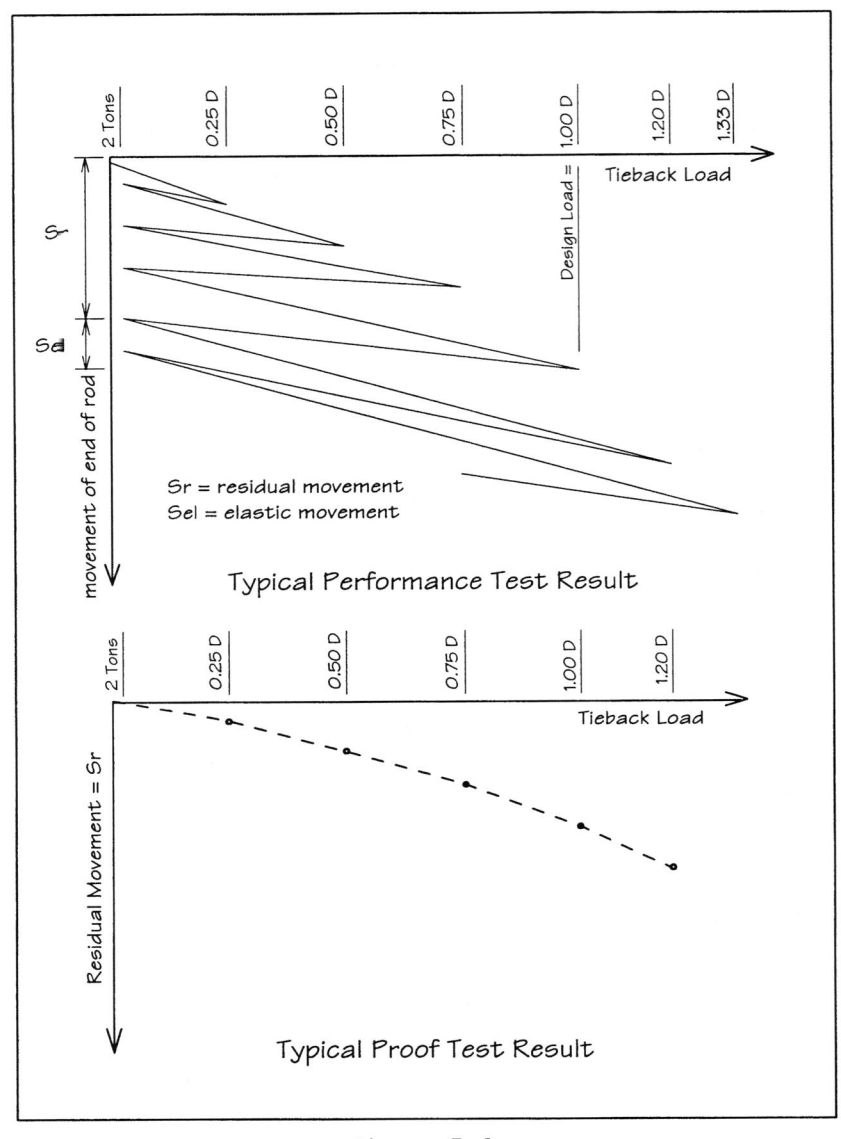

Figure 7-6
Examples of Tieback Pre-Testing

calibrated pressure gauge so that the jacking force can be set accurately. As a practical matter, a variation of about 5% of the load magnitude should be expected due to inaccuracies in the equipment. Load cells may be used to measure changes in load during the life of the structure, but are normally recommended only for more critical applications.

At the conclusion of each load cycle, the extension of the free end of the tendon should be measured with respect to a nearby benchmark so that readings are not affected by wall movement. Tendon movement should be recorded by a dial gauge sensitive to 0.001 inches (0.025mm). Extension of the tendon is plotted as a function of the applied load and compared to the theoretical elastic extension of the unbonded length. A successful test is one in which the actual extension is at least 80% of the theoretical extension, indicating that most of the load has been developed beyond the critical failure surface. For permanent anchors actual extension in excess of theoretical for the total of the un-bonded length plus one-half of the bond length could indicate anchorage failure.

Tendon extension exceeding theoretical for un-bonded length is considered to indicate movement of the bonded anchor. In granular soils particularly, some of this movement is likely elastic. In cohesive soils, the test load should be held for a specified time period to determine whether the bonded anchor is creeping. Creep movement during the test can be extrapolated over the service life of the structure to determine acceptability. Tiebacks which fail test procedures may still be used in the completed structure if down-rated to suitably reduced capacities.

7.4.3 Other Testing. Additional tests can be performed to establish benchmarks for performance of the anchors:

1. Performance Testing. A more elaborate version of the proof test with additional load increments, load cycles, and longer hold periods. Performance tests provide a detailed demonstration of anchor capacity and creep tendencies and are typically performed on about 10% of the production tiebacks. A typical plot of performance test results is shown in Figure 7-6.

2. Pre-production Testing. Sometimes carried out by contractors to confirm design assumptions. The engineer may require testing in special circumstances where there is concern about the suitability of tiebacks or the factors of safety which can be attained.

3. Tests to Failure. Tiebacks may be tested to failure to provide information on ultimate anchor capacity. Such testing is expensive because the structure must be reinforced to support potentially large loads. Although tests to failure can be useful for informational purposes, they may be difficult to justify from a cost standpoint.

4. **Lift-off Tests.** Recommended for most installations, particularly for permanent walls. It is appropriate to reposition the jack on a selected number of tiebacks to determine long-term creep tendencies. Lift-off tests help to verify design assumptions in that load distribution patterns may be identified for the entire wall. Such testing is usually performed at random on about 5% to 10% of the production anchors.

5. **Instrumentation.** It is recommended that survey data be accumulated on all tieback systems to assess the performance of the structure during and after excavation. A simple observational program consisting of bi-weekly transit surveys can provide valuable feedback about the validity of design assumptions. Such a program may help avert difficulties if unexpected wall movements are discerned early to allow modification of the excavation procedure. More elaborate measurements are maybe warranted if the installation is critical in nature. Examples include: load cells to monitor variations in anchor load; rod extensometers buried in the tieback drill holes to measure elongation; and inclinometer readings taken behind the support structure to obtain horizontal displacement profiles.

7.5 Summary

7.5.1 Tieback Anchor Advantages. The "as installed" capacity of individual tiebacks is demonstrated by proof testing during installation. This constitutes a significant level of quality control. Proof testing verifies the tieback design and provides data to modify it. Tiebacks can be added or deleted as required by test results.

Tieback systems offer the following advantages over conventional internal bracing:

1. Effects of temperature change are less pronounced with tiebacks than with internal bracing.

2. Frequently, bracing is removed and replaced to facilitate interior construction. Wall flexure occurs with strut removal and results in wall deflection. Tiebacks do not need to be removed. Tiebacks can be detensioned in stages in response to the installation of permanent supports.

3. The procedure of testing tiebacks and locking them off at a minimum of 75% of design load minimizes slack in the wall system and associated movement.

7.5.2 Difficulties with Tiebacks. As with any ground support procedure, difficulties can be experienced with tiebacks. The list below summarizes potential difficulties which should be addressed as part of a comprehensive design:

114

1. Mass Movement Beyond Anchorage. This would derive from an overall stability problem. The design should analyze all potential failure surfaces. Attention should be paid to over-consolidated soils with high residual lateral stresses and to situations where soft, cohesive soil is present below the excavation.

2. Anchor Creep. Time-delayed strain may not be readily apparent during short duration tests. Creep is encountered in plastic strain-softening soils and material subject to progressive strength reduction, such as fissured or slickensided clays.

3. Bearing Failure Beneath Structural Wall Members. The vertical components of anchor load must not be permitted to cause downward movement of the retaining wall. This may occur where wall elements are founded on rock above subgrade with potentially weak or weathered zones. When vertical members bear on soft soils, soil bearing capacity may be decreased by interior construction operations.

4. Corrosion. High strength steel tendons and associated gripping devices are subject to accelerated corrosion. Tendon corrosion may be accentuated by high levels of bending stress in the tendons immediately behind the wall. The longer the intended life span of a tieback, the more comprehensive its protection system must be.

5. Installation Methods. It is important to restrict certain drilling and grouting methods if there is potential for damage to existing structures. The contractor should be held accountable for any such damage and required to alter this installation method if it occurs.

7.6   References

7-1   Post Tensioning Institute (1980). *Recommendations for Prestressed Rock and Soil Anchors*. 1st Ed., Phoenix, AZ.

7-2   Schnabel, H. (1982). *Tiebacks in Foundation Engineering and Construction*. McGraw-Hill Book Company, New York, NY.

7-3   McMahon, D.R., Ed. (1986). "Tiebacks for Bulkheads," *Geotechnical Special Publication No. 4*, ASCE, New York, NY.

7-4   Goldberg, D.T., Jaworski, W. E., and Gordon, M. D. (1976). "Lateral Support Systems and Underpinning," *Report No. FHWA-RD-75-130*, Federal Highway Administration, Office of Research and Development, Washington, D.C., Vol. 3, 155-266.

7-5   Peck, R.B., Hanson W. E., and Thornburn, T. H. (1974). *Foundation Engineering*, 2nd Ed., John Wiley and Sons, New York, NY.

7-6   Ostermayer, H. (1974). "Construction, Carrying Behavior and Creep Characteristics of Ground Anchors," *Proceedings*, Conference on Diaphragm Walls and Anchorages, Institution of Civil Engineers, London, Session 5, 1-12.

7-7   Stille, H. (1976). *Behavior of Anchored Sheet Pile Walls*. Royal Institute of Technology, Department of Soil and Rock Mechanics, Stockholm, Sweden.

7-8   Nicholson, P.J., Uranowski, D. D., and Wycliffe-Jones, P. T. (1982). "Permanent Ground Anchors: Nicholson Design Criteria," *Report No. FHWA-RD-81-151*, U.S. Department of Transportation, Washington, D.C.

7-9   Otta, L., Pantueck, M., and Goughnour, R. R. (1982). "Permanent Ground Anchors: Stump Design Criteria," *Report No. FHWA-RD-81-152*, U.S. Department of Transportation, Washington, D.C.

7-10   Pfister, P., Evers, G., Guillard, M., and Davidson, R. (1982) "Permanent Ground Anchors: Soletanche Design Criteria," *Report No. FHWA-RD-81-150*, U.S. Department of Transportation, Washington, D.C.

CHAPTER 8: SYSTEM CONSTRUCTION AND PERFORMANCE

8.1 Introduction

This chapter is concerned with those aspects of construction of the support system which are of interest to the professional civil engineer who has planned, designed or specified the system. It is presented roughly in chronological order of the engineering activities which are involved in the construction, from the assignment of tasks and responsibilities, to difficulties which should be anticipated, necessary protection of adjacent structures, form of the specifications, controls to be exercised on the installation, inspection functions, evidence of developing difficulties, provisions for job review, and taking required action.

8.1.1 Responsibility for Support System Design. The decision of the owner and his engineer as to responsibility for design of the braced excavation depends upon the circumstances of the project, applicable building codes, and the contractual relation between the owner, the engineer, the contractor, and applicable building codes. Fundamental concerns are: constructability of the permanent structure, protection of life and property during construction, the interface with contiguous contracts, and potential liability of those involved, including the interests of impacted third parties and the public.

Some municipal building codes may require that the engineer-of-record for the permanent structure also design the earth retaining structure. This design may be developed as a preliminary design for cost estimation purposes in advance of construction, and later as a detailed final scheme, coordinated with the contractor's proposed construction methods. In some instances, public agencies may impose guidelines for the support system design.

8.1.2 Liability and Insurance. In case of a damage claim, the division of liability is influenced by the relative weight of insurance coverage of the parties. If project insurance coverage is maintained by the client, the owner's engineer has the greater responsibility for examining the scheme for the earth retaining structures in the light of potential for damage claims. Where the insurance is provided by the contractor, the owner's engineer, while still concerned with safety, may be somewhat relieved of the primary responsibility for damage. Specifications routinely state that the contractor is to assume responsibility and liability for all injury to persons or damage to property resulting from installation of the support system. Despite such statements, if the owner's engineer has taken a role in design and construction inspection, the engineer will share responsibility with the contractor to some degree. While the engineer may attempt protection by disclaimers in the specifications, the best defense is in the professional and technical quality of engineering services provided.

117

8.2 Performance Difficulties to be Considered

A number of problems are associated with performance of the overall excavation support system. These difficulties can lead to injury, damage, delay, unexpected expense, controversy, or adverse public reaction even in absence of apparent damage. An excellent summary of such practical considerations is given by in Reference 8-1.

8.2.1 Deficiencies in the Support System. Usually not the result of a gross error in establishing loading conditions or selecting main structural members, these deficiencies involve a failure to recognize adverse ground conditions or more subtle difficulties wherein poor details, secondary stresses or failures of workmanship may produce an undesirable effect. They include the following factors:

1.   Unsatisfactory Connection Details may result from failure to allow for the irregularities of retaining structure installation or lack of tight contact for the transfer of load from wall to bracing system. The critical requirement is to provide a connection suitable to support the design loadings which will serve adequately even with inaccurate positioning of structural elements, irregular contacts and flexing of the wall.

2   Secondary Stresses in Braces may result in unsatisfactory slenderness ratios which amplify dead load bending, or by eccentric load application, temperature changes, or construction activities. Excessive flexural stresses can lead to progressively increasing deformation. Buckling of compression members can initiate progressive failure if excessive deflection of crucial members causes load transfer through a stiff wale system to other members which are then overloaded and become unstable.

3.   Inadequate Passive Resistance. Inadequate berm sections, berm deterioration or decrease of support at subgrade level caused by disturbance during construction can reduce the sheeting support. If resistance of an interior berm is critical, the designer must estimate the portion of full passive resistance that will be available under actual construction conditions. The amount of movement necessary to mobilize the required resistance should be evaluated.

4.   Total System Instability may arise from oversight or inadequate investigation of the possibility of base failure. While it may be accentuated by construction operations, instability often results from an unfavorable subsoil profile wherein deep-seated movements are inherent based on the excavation geometry. A proper stability evaluation should be based on a competent geotechnical investigation.

118

8.2.2 Excessive Movement. Braced excavations normally will experience movement. Excessive movement can result from structural difficulties, construction procedures, or unfavorable subsoil conditions. The following factors are frequently involved:

1. Spacing of Excavation Support Elements. Excessive vertical spacing of bracing tiers or horizontal spacing between soldier piles may result in spans that are too great at critical stages. Such conditions tend to promote movements along the sides of the open excavation and/or beneath the adjacent subgrade.

2. Inadequate Support Stiffness results from excessive member slenderness, inattention to blocking and wedging details, or lack of posting, ties, stiffener plates, etc. Deficiencies of detail can combine to produce slack in the system which can lead to continuing gradual movements even after excavation is complete.

3. Wall Stiffness Requirements. Various wall elements can be used, ranging from relatively flexible timber sheeting, through soldier piles and lagging and steel sheet piles, to relatively rigid concrete diaphragm walls. Wall stiffness can play a critical role in controlling ground movement, especially for soil profiles involving very soft to medium clays. The relationships among wall movement, section stiffness of the wall, vertical spans, and factor of safety against base heave in clay are discussed in Chapter 4.

4. Inappropriate Excavation Procedures. Prime examples in a soldier pile and lagging system are improper packing of the space behind lagging and delay in the installation of lagging so that sloughing or running of the retained soil occurs. Over-zealous efforts to deepen the excavation before bracing is placed or excessive trenching and ditching at the toe of the wall can decrease interior resistance and promote inward movement of the wall near excavation level. In the case of slurry wall construction, excavation techniques are critical factors for control of movement. Panel length and depth, guide wall position, and slurry characteristics should be selected with consideration of their effect on system stiffness.

8.2.3 Unsatisfactory Water Control. Difficulties arise when the dewatering system is inadequately designed at the outset or is not operated in response to conditions encountered in the field. Runoff of surface water outside the wall can aggravate the piping of soil through lagged soldier pile walls, at wall openings, or where the top of wall is below surrounding grade. Exterior settlement due to gradual loss of soil or loosening of the contact between the sheeting and retained material until there is a sudden slumping of the ground can result. Runoff can also interfere with some types of tieback installation. Troublesome loss of soil can be caused by leaks from utilities discharging a concentrated flow toward the wall. If

119

the wall forms an effective water barrier, leaking from utilities can increase pressures acting on the wall. Besides making earth moving difficult and costly, unsatisfactory water control in the interior of the excavation can disturb the interior subgrade and berms, and reduce passive resistance to sheeting movement.

8.2.4 <u>Construction Coordination</u>. Difficulties can result from divided responsibility or conflicting interests where subcontractors are operating at cross purposes. To the extent practicable, specifications should make provision against ambiguous or divided responsibility. For example, the earth moving contractor may attempt to remove a maximum amount of material, over-excavating below brace levels or over-steepening berm slopes. Dangerous stockpiling of spoil or materials near the top of wall or damage to support system elements by impact of excavation equipment are common events during excavation. Pile driving may inadvertently cause difficulties by penetrating the interior subgrade. In silt or silty fine sands, this can produce excess pore water pressures and quick conditions with loss of passive resistance at subgrade. Difficult driving of wall elements may lead to out-of-plumb or twisted soldier piles or jumped interlocks and misalignment of steel sheet piles. This complicates the challenge of providing unyielding support and, as the excavation is deepened, the threat of ground loss increases. Corrective measures and repairs may be difficult but are frequently necessary to correct what could otherwise constitute a weak link in the wall system.

8.2.5 <u>Avoiding Subgrade Disturbance</u>. Where disturbance of the subgrade could jeopardize the temporary or permanent structure, it may be advisable to control the equipment employed, methods of excavation, or other construction activities as the final subgrade level is exposed. Final excavation should be performed by methods which minimize disturbance. Such methods can include use of light bulldozers, overhead loaders, or hand work. It may be advisable to restrict use of heavy clamshell buckets or backhoes or the presence of heavy earth-hauling equipment on the final subgrade. Disturbance can be accentuated by the presence of water or inflow of seepage near the base of the excavation. Ordinarily, specifications should require advance drawdown below final subgrade. Conformance should be monitored by installation of piezometers or observation wells.

8.3 Integration with Permanent Interior Structures

After the interior permanent base slab has been constructed, risk of excavation instability is significantly reduced. As the permanent structure is constructed, the integrity of the excavation support system increases as its function is augmented by the permanent structure. During construction, support tiers should be removed only after the permanent structure, completed to that particular stage, can sustain the load transferred to it. In integrating the permanent and temporary support systems, the following factors should be taken into consideration:

1. Temporary braces should clear the zones to be occupied by the permanent floor system. The optimum location for a horizontal strut is usually directly above a permanent floor level. For a raker, the optimum location is just below the floor. The raker can be removed after the permanent floor can support the load (allowing forming and placement of the next wall lift and floor slab). Where interior bracing cannot be removed, provisions must be made to embed portions of the bracing within the wall, leaving a recess in the interior face of the wall within which the strut can be cut and the wall patched.

2. Temporary piles which must remain during interior construction (such as for support decking or equipment trestles) should be positioned to avoid conflict with permanent walls and columns. Portions penetrating the base slab can be left embedded in the slab with a recess for later patching and waterproofing. Upper sections may need to be boxed out (for subsequent pours of floor slabs) and later removed.

3. Procedures for removal of bracing and rebracing should be worked out in advance of construction and either prescribed by the engineer or proposed by the contractor and reviewed by the engineer. Where embedment of a temporary support element is necessary, it should be planned to fit properly with the encasing concrete. Sometimes the need for an unencumbered height for interior concrete pours may require increasing the unsupported span of the wall to a greater amount than previously experienced. In this case, loads on the next upper tier of braces remaining in place may be maximum and positive moments in the span of the wall may be more critical than those during initial installation. The design should take into account this possible scenario.

4. Position tolerances for the permanent structure and adjacent excavation support elements should be consistent with the accuracy that can reasonably be obtained during excavation support installation. Appropriate allowance for movement, displacement or distortion should be considered. Consideration should be given to the fact that the permanent structure can be placed to closer tolerances than can the temporary support system, the position of which is subject to difficulties of alignment and soil movement.

5. Redistribution of lateral earth loads in later stages of construction can be significant especially for deep excavations. Redistribution can result not only from removal of braces and rebracing but also from increases in water levels, the pattern of soil movements, and soil creep. Pressures on the permanent structure tend to increase at lower levels and decrease at upper levels compared to the temporary excavation loads (Reference 8-2).

6. Specifications should require that groundwater control continue as long as necessary to prevent flooding and to control uplift pressures (which could produce either unacceptable bending stresses in recently placed base slabs or an unbalanced total uplift force). There is some benefit to permitting groundwater levels to rise as soon as practicable after construction to test the watertightness of the permanent structure. However, control of this rise should be carefully staged with respect to permanent construction.

Removal of excavation support requires procedures which are pre-planned, tailored to fit the project, integrated into the construction, and clearly addressed in the specifications. Localized ground movements can result from holes made by removal of support elements. Vibratory extractors can minimize the size of hole made by removal of sheetpiling or soldier piles because the amount of soil clinging to the steel is decreased. This hole should be filled to the extent practicable. However, filling is often problematic because the hole closes as the piling is removed, producing a commensurate lateral movement. Where the potential effect on adjacent property is judged unacceptable, wall elements should be left in place. However, wall elements should be cut off and removed near ground surface to avoid formation of hard points at ground surface.

8.4   Pile Driving Problems

Excavation support design should consider the effects of pile driving for the interior structure or adjacent exterior areas. Pile driving can disturb and weaken silt or clay of soft to medium consistency such that shear strength is lowered and compressibility is increased. When driving large groups of displacement piles into saturated fine-grained soils, heaving of the subgrade surface can occur along with the other forms of disturbance. There may also be a tendency for lateral displacement of the soil outward from the pile group. Where piles are driven inside the excavation adjacent to berms, significant movements and reduction in passive resistance can result. Pile driving can cause an increase in pore water pressures in silt or clay or loose silty sand followed by ground settlement and downdrag forces on the piles. Poorly graded sands can be densified by pile driving. This densification can extend for some distance beyond the perimeter of the installed pile and can affect nearby structures.

Driving of sheet piles, especially when vibratory hammers are used, can induce settlement in loose and medium sands. Reference 8-3 presents case history information on the effects of vibratory hammer driving on the settlement of loose sand adjacent to a deep braced cut. Reference 8-4 summarizes field measurements and analyses of transient and permanent ground displacements resulting from vibratory hammer driving in loose to dense sand.

The following factors should be considered to minimize detrimental effects of pile driving:

1. Where pile support is obtained in a deep bearing stratum, detrimental effects can be decreased by predrilling holes for a portion of the pile length above the bearing layer. Soldier pile holes are frequently predrilled and thus require backfilling around the installed soldier pile. This backfill should consist of a cohesive material that can be trimmed readily but will remain stable during subsequent excavation. Lean dry concrete, soil cement or a bentonite-cement-sand mixture may be appropriate.

2. If predrilling for soldier piles is carried to a level below subgrade, the interior annular space should be filled with concrete to increase the resistance (horizontal and vertical) of the buried portion of the pile. Otherwise, the soldier pile may undergo additional lateral movement before developing the passive resistance of the soil below subgrade.

3. Installation of permanent foundation piles within a braced excavation requires careful planning to avoid interference with internal bracing crossing the foundation area. Piles can be driven at an early stage of excavation before bracing is installed utilizing a follower to bring the pile top to approximate cut-off level. If the soil conditions require a pile foundation for the permanent structure, the act of driving piles can have a detrimental effect on soil strength and stability. Selecting the best stage for pile installation will depend upon the type of soil, the plan for groundwater control, the presence of a bracing or tieback system, the need to support pile driving equipment and the comparative cost of excavation with or without the piles installed.

4. Installation of permanent foundation piles with a follower through soil eventually to be excavated should take into account the detrimental effect of holes created by the follower and the effect of soil disturbance on passive resistance at the base of the excavation. Backfilling of follower holes requires immediate attention and careful control.

5. Pile driving vibrations transmitted to adjacent buildings can cause annoyance, cracking and other damage. Consideration should be given to minimizing the effects of pile driving by predrilling. Restrictions on the time and hours of pile driving are often set by local regulations.

8.5 Problems of Rock in the Excavation

Excavations that extend into rock and require blasting can involve certain difficulties. The presence of rock will limit the choices of types of support systems and methods of construction.

The following factors should be considered in design and specifications:

1. Restraint of a rock face below a wall supporting overlying soil by rock bolts, shotcrete and wire mesh, or bracing from face to face may be necessary. Line drilling or presplitting or other smooth-wall blasting methods can minimize rock face stability problems. Sometimes, it is not prudent to blast to the designated excavation depth, but rather to conduct final rock removal using a hydraulic backhoe or comparable equipment. Measures to be taken depend on the jointing and quality of the rock. Unfavorable dip of rock joints downward into the cut or deep weathering may make it necessary to drill through rock to several feet below subgrade for installation of soldier piles or principal wall elements.

2. Optimum procedures for drilling and blasting and blast hole patterns will be dictated by the character of rock to be excavated and geometric features of the excavation. All blasting activities should conform to industry safety practices and any applicable ordinance. Acceptable procedures should be set forth in the specifications with provisions for modification based on results of initial trial blasts. In developed areas, blasting mats are generally necessary for control of fly rock.

3. Pre-blasting condition surveys of adjacent structures and continuous monitoring may be necessary. Because cosmetic damage to adjacent structures is difficult to prevent and blasting is inherently dangerous, consideration should be given to alternative means of rock removal. Alternative means include hydraulic fracturing with rock jacks and pneumatic fracturing by impact devices. Cost studies of rock excavation may show alternative methods to be more economical than blasting. In some cases, the upper portion of bedrock will be weathered and jointed to the extent that a rock rake or ripper can dislodge much of the material in sizes suitable for removal by backhoe or front-end loader.

8.6 Protection of Adjacent Structures

If excavation might could conceivably cause damage to adjacent structures, owners will often attribute damage to the excavation regardless of whether or not movement or damage has actually occurred. To provide documentation to counter such a contention, a careful preconstruction survey of adjacent structures should be conducted. Buildings should be resurveyed from time to time, particularly when movements are possible or when vibration or blasting is likely.

A general assessment of the need for underpinning or protection of adjacent structures should be made in the design stage. Uncertainties and the state of the art make such preconstruction decisions highly judgmental. Actual movement must be continually monitored and re-evaluated during construction. Specification provisions for protection of existing adjacent structures may range from stipulating the contractor's responsibility to prevent damage to inclusion of an engineer-designed underpinning system. "Underpinning" is taken to mean the introduction of new substructure units beneath existing adjacent structures for the purpose of lowering the support level. "Protection" of the adjacent structure implies the restraint of movement by providing a stiff, stable support system or by some form of soil stabilization.

The owner and the engineer must decide on the requirements to be specified (underpinning versus protection and method versus performance specification). Factors influencing these choices are:

1. <u>Geometric Relationships</u>. Often, the concept of an "influence line" sloping upward from the excavation subgrade is utilized to assess the need for underpinning. For example, it may be assumed that adjacent structures which are founded inside a wedge bounded by a slope of 1 horizontal to 2 vertical from the base of excavation should be considered for underpinning or protection. For foundations outside that line the need for underpinning can more likely be avoided although some protective arrangements may be required. In unfavorable subsoil conditions or with particularly sensitive structures, a conservative choice is made within the range of such influence lines. Alternatively, movement associated with the excavation may be estimated from case history data, such as those summarized in Reference 8-5, and likely effects on adjacent structures assessed to determine underpinning/protection requirements.

2. <u>Subsurface Conditions</u>. Unfavorable subsurface conditions include loose sands or non-plastic silts, soft silts and clays, fill, organic soils, and high groundwater. Favorable subsoil conditions include compact sands and gravels, stiff silts and clays, rock and rock-like materials, and low groundwater. Especially troublesome loss of ground can be expected in a lagged or non-continuous wall in cohesionless silt or sand with a high water table. In the absence of systematic exterior drawdown, a run of material through wall openings can cause movements of the surrounding ground.

3. <u>Character of Adjacent Structures</u>. Appropriate protective measures are dictated by the age, importance, use, physical characteristics, and

economic value of the adjacent structure. Modern concrete or steel frame buildings with mat foundations, grade beams, or continuous footing systems are probably the building types least vulnerable to damage by movement. Most vulnerable are old brick or masonry buildings with essentially no continuity and little ability to resist horizontal extension strains. Obviously, the value or importance of the structure must be balanced against the cost of protection. Occasionally, it is expedient for the owner of the new construction to purchase a vulnerable adjacent building. In other cases where damage is expected to be superficial, it may pay to agree to necessary repair or cosmetic treatment rather than to undertake a costly underpinning or protection scheme.

4.   Nature of the Support System. In some cases, it may be desirable and practicable for the owner's engineer to dictate the type of excavation support system that will serve for protection either as a supplement to or in lieu of underpinning. Use of special walls, such as slurry walls, tangent piles, or pit walls with proper bracing can obviate the need for conventional underpinning. To afford maximum protection, restrictions on the spacing of the wall units or bracing tiers and on construction procedures may be necessary. Factors such as the presence of utilities, space available for the wall, integration with the permanent structure, and cost factors influence the selection of the support system for protective purposes.

The owner of new construction may specify either directly or through the owner's engineer certain protective measures. Alternatively, protection may be left to the choice and responsibility of the contractor. Factors that influence the approach include: the attitude of the abutting owner; the length of time adjacent buildings are exposed to damage during critical stages of excavation; availability of space in or around the adjacent structure for installation of protective measures; the disruptions such installation may cause; and local codes and regulations relating to responsibility for protection. Design of underpinning or alternative protective schemes should be as detailed as the design for permanent construction.

Specifications relating to the protection of adjacent structures should cover the following:

1.   The basic requirements to be satisfied by the underpinning or other specified measures and the criteria of acceptability that would be applied to the contractor's proposed alternative scheme. If protection of adjacent structures will rely on stiffness of the support system, specific requirements toward that end should be described. All information on the adjacent structure should be made available. Any requirements such as limitations on movement, preservation of the building function, assignment of available work space, etc., should be clearly set forth.

2. Arrangements for pre-survey and monitoring of adjacent buildings, and contingency measures to be implemented in case of unsatisfactory should be clearly described.

3. The sequence of underpinning, if required, with respect to the nearby excavation should be stipulated. Underpinning work should be completed prior to any adjacent excavation below the level of the existing structure foundations.

4. Should the work on underpinning or other protective measures reveal unforeseen foundation conditions, difficult dewatering, distress of the structure or change in subsurface conditions, then revision of the plans may be necessary. Such revisions require timely inspection and monitoring of the work and the structure behavior.

8.7 Specifying the Excavation Support System

As with other temporary construction features, the support system specifications may be developed as "method" or "performance" type. In most elaborate form the method specification includes a fully designed system as complete as for permanent construction with requirements for staging, methods, and procedures which tightly stipulate the contractor's options. At the other extreme, a minimum performance specification would consist of merely the functions that the support system must fulfill, such as, permitting the excavation to be conducted safely in the dry while limiting ground movements and protecting adjacent structures.

As a general rule, framing specifications to achieve an apparent initial economy in the excavation by omitting or de-emphasizing technical requirements may be ill-advised and finally uneconomical. Inclusion in contract documents of a complete support system design can promote more equitable competitive bidding. Bidders may not include the range of contingencies necessary for their tentative design and the specified system can be used as a point of departure for bidder alternatives. On the other hand, qualified specialist contractors may offer unique support systems which accomplish a specified result and affect significant savings for an owner. A method specification will limit this sort of contractor initiative. The choice between "method" or "performance" specifications is thus often best made with knowledge of the prospective bidders and their capabilities.

The choice between method or performance specification depends on the scale of excavation, characteristics of subsoils and nearby structures, qualification of expected bidders and the anticipated quality of the construction control and inspection. The owner should consider providing all or part of the excavation support design in the contract if the following conditions pertain:

1. There is a lack of precedence and/or high risk of serious consequences if unsatisfactory performance occurs.

2. The earth retaining structure is to become an integral part of the permanent construction or will be an important element in the protection of adjacent buildings.

3. Elaborate analyses of loading, stability or soil-structure interaction are necessary and the owner's engineer is well qualified for such studies.

4. Interface conditions with adjacent contracts must be resolved at an early date.

5. Building code provisions made a contract design expedient or necessary.

Assuming that the owner and the owner's engineer agree that economy and efficiency would be served by a support system design in the contract, an essential additional requirement is to stipulate the criteria which will be applied to the contractor's proposed alternative to the contract design. This alternative may be submitted in the form of a value engineering proposal which may offer potential savings to the owner but which may fail to meet certain requirements which are intrinsic (but perhaps not obvious) in the contract design. In fairness to all bidders, the criteria surrounding acceptance of other support schemes must be clearly stated. It is important to include any special requirements that must be satisfied with respect to the permanent interior construction, the time of performance, or methods which will be imposed on any alternative scheme for protecting existing adjacent structures.

Where appropriate, the owner may leave planning and design of the earth retaining structure entirely to the contractor, only stipulating requirements for protection of third party interests and limitations imposed by the presence of the permanent structure. It is a common approach for routine excavation jobs. This course of action might be chosen where adjacent sites are not sensitive to job operations and third party involvement is minimal or where a selected contractor has special knowledge of subsurface conditions or special capabilities in construction. The following factors should be considered in formulating the specification:

1. Stipulating lateral pressure diagrams as design criteria will at least provide a common basis for bid and a standard for reviewing the adequacy of the contractor's proposed support system. These pressure diagrams should include special surcharge loadings, street deck loads, or adjacent building loads that have to be resisted. It could be helpful to set forth acceptable methods of computing brace loads and determining stresses in the main bracing members.

2.   Necessary limitations on the contractor's selection of materials, sheeting and bracing types, and geometry may be included, as well as limitations on the vertical spacing of piers or braces or restriction on the depth of excavation below the position of a brace yet to be installed.

3.   A performance specification should include criteria to be met by the excavation support system, such as limitations on lateral movement and groundwater drawdown or restrictions on the time intervals of critical importance. Measures that are required or are permissible for protection of adjacent structures should be defined.

In some instances, the contractor's technical staff will prepare working drawings to supplement the specified design, or will prepare design and working drawings based on stipulated criteria. Alternatively, the contractor may make a completely independent evaluation of subsurface conditions and develop the excavation support scheme.   Thus, to varying degrees, the contractor's technical staff and his consultants must supplement the work of the owner's engineer in achieving satisfactory support of excavation. The working drawings and calculations should be stamped and signed by a registered professional civil engineer.

Specifications should cover in detail the requirements for the contractor's working drawings, procedures for their submittal and the necessary calculations and design details for the owner's engineer's review.   This review normally is to confirm that contract requirements and design criteria are satisfied and that conflict with other related work is avoided. The purpose of the review is not to check all details that are within the contractor's options. Nevertheless, when a reviewer notes that the working drawings conform to criteria, the reviewer may inherit responsible to the degree that the criteria are appropriate to the work.   During construction of features covered by the working drawings, it is the contractor's responsibility to provide what is shown on the drawings as if they were part of the contract.   The contractor's field staff are responsible for installation of the earth retaining structure in accordance with design and specification provisions.

8.8   Inspection and Observations

Field inspection forces may be staffed by the owner's engineer, the owner, or by organizations specializing in such work.   The choice depends on the owner's policy, the qualifications of his engineer, the project size, scope, or location, or local code requirements. Specifications should clarify the division of authority and responsibility.   To facilitate control and to evaluate conformance with contract requirements, a program of field observations, possibly including instrumentation, may be set forth in the specification.   In planning an observation program, the following factors should be considered:

1. Specifications should clearly state the objectives and features of this program, including measuring points to be established, observations required and instruments to be installed. Specifications should delineate duties for making observations.

2. Normally, the owner will provide control surveys for the project while the contractor will undertake day-to-day construction surveys. Routine measurements in the observation program are ordinarily made by the contractor's field party while more specialized observations and management of complex instrumentation are better assigned to the engineer or a collaborating specialist.

3. Where the owner's engineer is responsible for field inspection, his representatives should participate in the observation program. Data should be recorded and evaluated on a timely basis so that it can be of immediate use for control purposes.

8.8.1 Resident Engineer and Field Inspector. When a resident engineer represents the owner or the owner's engineer in on-site inspection, the field inspector's functions are to observe the installation, record observations in a consistent format, report with routine written records, and notify the resident engineer of occurrences that may influence performance or costs. Specifications should require the contractor to provide the inspector with necessary access to make the required observations, particularly when this could delay construction, inconvenience contractor personnel, or require their assistance.

The extent and frequency of inspection should be defined in the specifications. Inspection should be frequent during critical stages including installation of sheeting and bracing elements. The inspector should indicate those features shown on the drawings confirmed by direct observations as well as conformance or variance of the actual installation with the drawings. The inspector should inform the resident engineer of details of the work not covered by routine reports such as: discussions with the contractor and other parties, agreements and understandings, special notes regarding equipment or organization, labor conditions, weather, and changes in subsurface conditions.

8.8.2 Record Keeping. Support system installation records should document as-built information developed during construction by the inspector, resident engineer, and contractor. Appropriate records generally take the following forms:

1. Daily reports tabulating work accomplished, noting unsatisfactory or partially completed items, corrections of previously reported unsatisfactory items, instructions given or received, delays encountered, and the statistics of job operation. When a field observation program is involved, it is

desirable to include a special schedule for recording or summarizing observations and status of instrumentation.

2. A daily diary and minutes of all job meetings. The resident engineer reports job progress on a weekly basis to give an update on actual versus planned progress, overall physical conditions for that week, job accomplishments, and summarizes current or potential job problems.

3. A report to summarize data and observations to evaluate support structure performance, noting any significant departures from design assumptions and items which could impact cost or quantities.

8.8.3 Provisions for Job Review and Taking Required Action. Construction records should contain the facts and details necessary to carry out an engineering evaluation of the integrity of the support structure and to provide a basis of payment to the contractor. Specifications should stipulate provisions for coping with unplanned events such as cave-ins, excessive yielding, structural collapse, threatened flooding or subsurface conditions differing from those anticipated. The contractor should be given the authority and responsibility to maintain safety and protect property. Procedures for revising design criteria, plans, and specifications should be provided for both routine and emergency circumstances.

8.9 Response to Construction Difficulties

In design of a major braced or tied-back excavation, the anticipated vertical and horizontal movements of the retained ground should be evaluated from available case history data such as those in Reference 8-5. The program for construction control and inspection should be planned to identify any departures from expected performance and to provide for corrective measures. Some simple means should be provided for determining the basic vertical and horizontal components of movement of the cofferdam, retained soils and adjacent structures. Contract drawings should give an indication of desired locations of observation points and specifications should stipulate a frequency of readings. When widespread drawdown, vibration or construction activities could cause general settlement, surface benchmarks should be established beyond the affected area. Specifications may stipulate that special instrumentation be installed, such as inclinometers or various sorts of structural strain gauges.

8.9.1 Difficulties with Installation. Conditions that could lead to difficulties often can be identified by direct visual evidence or indications that can be obtained by simple measurements. These conditions include wall elements that have moved out of plumb, bulging of lagging, connections where gaps are present between elements which should be in contact, or obvious eccentricity of load transfer. Any appearance of instability such as excessive sagging, deflection, twisting or distortion of structural members must be identified. In many instances, it is difficult to ascertain whether the apparent distortion is a product of the initial installation or changes caused by excavation operations. It is important that damage to support units, such as jumping of sheet pile interlocks or twisting of soldier piles as yet unexposed by the excavation be identified in advance when possible. When visual evidence indicates possible difficulty, measures should be planned to counteract the difficulties in a timely fashion.

8.9.2 Loads in the Support System. In critical work involving protection of adjacent properties, preloading of compression braces or pretesting of tiebacks can be essential, as discussed in Chapters 6 and 7. Preload can be verified in the field by load cells, gauge readings of hydraulic jacks, torque wrenches, or by various forms of strain measurements (converted to stress and then to load). Provision should be made for emergency procedures for reinforcement or addition of braces if measurements or other performance evidence indicate potentially unsafe strut or tie loads.

The simplest methods of determining strut loads employ devices to measure change of length over known distances. Temperature corrections should be made to distinguish the sources of stress increase. For consistence, it is desirable to obtain measurements at the same time of day or, more importantly, at as close as possible to the same temperature on any day to minimize temperature effects. Methods for maintaining a fairly constant strut temperature include water spraying, misting, painting with heat-reflective paint or use of a deck or covering to shade the support system.

If observed movements or measured loads indicate that supplementary bracing is needed, struts may be added and secondary bracing and posting increased as necessary to prevent buckling. The installation of tiebacks to arrest movement may be beneficial. Stiffening of structural members by welding on plates should be pre-planned as a contingency procedure and subjected to a careful structural analysis. Without proper analysis and guidance in the specifications, this practice has could lead to weakening of supports and detrimental eccentric loadings.

8.9.3 Conditions of the Excavation Interior. Difficulties often involve unsatisfactory water control, such as softening or heave of subgrade soils, elevated piezometric level in observation wells, or inflow of seeping water. Apart from the conclusion that the dewatering system may be inadequate, other factors may be contributing which should be investigated. These include: damage or breaks in the buried portion of a watertight wall; leakage from nearby utilities; the concentration of shallow seepage in some convenient flow channel such as fill in an old stream or utility trench; or the presence of unexpected pervious strata beneath subgrade which are not tapped by the dewatering system. Where dewatering is necessary, the specifications should require the contractor to install an appropriate set of observation wells, sealed against inflow of extraneous surface water. They should be utilized to identify impending difficulties and to monitor conformance with dewatering requirements.

8.9.4 Response to Field Observations. If observations indicate an unsafe condition is developing, revisions to the construction procedure or basic design may be necessary. Where settlement of the surrounding ground is observed without comparable horizontal movement of the sheeting, consolidation from drawdown or as a result of pile driving or other sources of the vibration could be the cause. In the event of distress to the excavation support systems or adjacent structures, emergency measures may become necessary. It may become necessary to suspend excavation activities to permit accelerated installation of planned supports or the addition of supplementary supports. Controlled flooding might be necessary to offset erosion or piping of soil by water flow into the excavation. Partial backfilling on a selected basis could offset the damaging effects of overexcavation. Berm dimensions could be increased and ground water levels within the berms further lowered in order to increase support of the buried portion of the sheeting. If difficulties at an early stage dictate strengthening the support system, consideration should be given to redesign of the support system.

8.10 References

8-1   Peck, R. B. (1976). "Performance of Lateral Support System," *Proceedings,* Geotechnical Lecture Series on Lateral Earth Pressure, Boston Society of Civil Engineers, Boston, Massachusetts.

8-2   Gould, J. P. (1970). "Lateral Pressures on Rigid Permanent Structures," *Proceedings,* ASCE Specialty Conference on Lateral Stresses in the Ground and Design of Earth Retaining Structures, Cornell University, Ithaca, New York.

8-3   O'Rourke, T. D. (1989). "Ground Movements Caused by Braced Excavations," *Journal of the Geotechnical Engineering Division,* ASCE, 107(GT9), 1159-1178.

8-4   Clough, G. W. and G. L. Chameau (1980). "Measured Effects of Vibratory Sheet-Pile Driving," *Journal of the Geotechnical Engineering Division,* ASCE, 106(GT10), 1081-1099.

8-5   Clough, G. W. and O'Rourke, T.D. (1990). "Construction Induced Movements of Insitu Walls," *Design and Performance of Earth Retaining Structures,* edited by P. C. Lambe and L. A. Hansen, ASCE, New York, 439-470.

INDEX

135

tem 28, 33; monitoring changes 38-39; pumping capacity 27; soldier pile wall installation 76; subsoil conditions 28-29. *See also* Surface water control; Water control

Groundwater levels 9-10; monitoring 39

Groundwater loading 59-61

Grouting; support wall formation by 79-80

Gunite; as lagging between soldier piles 80

H

Heave 36, 37, 47

I

In-situ strength testing 11

Inclinometers 17-18

Insurance 117

Interior berms 91-92, 93, 118

Internal berms; stability analysis 47

Internal support systems 84; bracing 85-91; components 85-92; construction difficulties 131-133; construction and performance 117-133; construction requirements 98-100; deficiencies 118-120; design 117; design loads 92-93; field inspection and observation 129-131, 133; integration with permanent interior structures 120-122; interior berms 91-92, 93, 118; liability and insurance 117; loads 92-94, 95; performance difficulties 118-120; permanent floor systems 91; pile driving problems 122-124; prefabricated systems 92; preloading 85, 87-98; protecting adjacent structures 124-127; ring beams 92; specifica-

tions 98-100, 127-129; stability 93-94; structural design criteria 94-97; timber material 92; top-down construction 91; wale support 92. *See also* Support walls

J

Jelinek and Ostermayer method; stability analysis 45

L

Lagging walls 74-77; properties 70; standards and specifications 72

Liability 117

Lift-off tests; tieback support systems 114

Loading of supported excavations 132; bracing 40, 92-94, 95; checklist 50; design loads 59, 81-82, 92-93; earth loading 40, 42, 51-59; earthquake loading 56, 58, 62-63; frost action 61; groundwater loading 59-61; internal support system 92-94, 95; redistribution during construction 121; safety factors in design 65-66; shear strength 63-65; temperature fluctuation on braces 61-62

Lumber; trench bracing 92

M

Micro-piles 80

Modified Bishop procedure; stability analysis 45

Monitoring; groundwater levels 39

Morgenstern and Price method; stability analysis 45

O

Ordinary method of slices; stability analysis 45

Owner; defined 5